D0875457

PRENTICE-HALL

FOUNDATIONS OF DEVELOPMENTAL BIOLOGY SERIES

Clement L. Markert, Editor

CANCER

A Problem of Developmental Biology

G. Barry Pierce
Robert Shikes
Louis M. Fink

University of Colorado Medical Center

PRENTICE-HALL, INC., Englewood Cliffs, New Jersey 07632

Library of Congress Cataloging in Publication Data

Pierce, Gordon Barry, (date)
    Cancer: a problem of developmental biology.

    Bibliography: p. 200
    Includes index.
    1. Cancer.   2. Developmental biology.
3. Carcinogenesis.   I. Shikes, Robert, joint author.
II. Fink, Louis Maier, (date) joint author.
III. Title.   [DNLM: 1. Neoplasms—Etiology.
2. Neoplasms—Pathology.   QZ202.3 P616c]
RC261.P538   616.9'94'07   78-72
ISBN 0-13-113373-X

PRENTICE-HALL
FOUNDATIONS OF DEVELOPMENTAL BIOLOGY SERIES

Printed in the United States of America

10   9   8   7   6   5   4   3   2   1

PRENTICE-HALL INTERNATIONAL, INC., *London*
PRENTICE-HALL OF AUSTRALIA PTY. LIMITED, *Sydney*
PRENTICE-HALL OF CANADA, LTD., *Toronto*
PRENTICE-HALL OF INDIA PRIVATE LIMITED, *New Delhi*
PRENTICE-HALL OF JAPAN, INC., *Tokyo*
PRENTICE-HALL OF SOUTHEAST ASIA PTE. LTD., *Singapore*
WHITEHALL BOOKS LIMITED, *Wellington, New Zealand*

Foundations of DEVELOPMENTAL BIOLOGY

Cancer represents a grave threat to human beings and, consequently, has been the subject of investigation in several disciplines of biology for many years. Although we now have accumulated vast amounts of information about cancer, we still are unable to formulate a generally acceptable and useful definition of the malignant cancer cell. Many disciplines of biology, genetics, biochemistry, virology, and immunology, to name but a few, have laid claim to cancer as an abnormal manifestation of biological phenomena within the purview of their particular disciplines. These claims have merit, but for several years cancers, or more generally neoplasms, increasingly have been appreciated from the point of view of developmental biology.

Neoplasms have been defined by developmental biologists as diseases of cell differentiation stemming from the misprogramming of normal gene function. In this monograph, Dr. Pierce and his colleagues present cancer as an abnormal manifestation of the processes of cell differentiation and of embryonic development. Their presentation contributes significantly to a definition of the malignant cell and to an understanding of its origin and growth within the afflicted individual. With incisive analysis and lucid presentation, the authors have sharpened the questions that may be asked by students at all levels in their study of neoplastic or cancerous development. The authors have brought to bear on this problem years of experience in working on the pathology of cancer and a lively appreciation of the principles of developmental biology which surely encompass the abnormal cell differentiation we know as cancer.

CLEMENT L. MARKERT

122416

Contents

Preface

The stimulus to write this book came from several sources. My studies utilizing developmental approaches to neoplasms yielded information about the cellular composition and interactions of cells of tumors that was incompatible with some of the traditional dogmas of cancer. Few oncologists are aware of the cellular heterogeneity of tumors and as a result few have even thought about the interactions of cells in tumors. We have all been steeped in the cellular concept of neoplasms, and have taken for granted that a cancer is composed of malignant cells only. This is wrong. Hopefully, this book will set the record straight. From the standpoint of cellular interaction, the data are so woefully incomplete that it is only possible to show how important the subject is and to explode the old dogma that tumor cells are uncontrolled. They are controlled. It is important to know that, too!

The final stimulus came from teaching pathology to students of zoology at the Universities of Michigan and Colorado and to graduate students in the Department of Pathology at the University of Colorado Medical Center. One day a graduate student told me that he had come to the department with an interest in disease. The department had taught him to be a good scientist, but had not provided the desired exposure to disease. Would we please rectify this problem? From this experience I learned that if a student says he is interested in something, he probably is. Furthermore, it was apparent that students have a curiosity and hunger for information about the abnormalities of biology. This book, written for graduate students, is about one of the major abnormalities. Hopefully, as the student learns about cancer, he may be able to utilize cancers as easily manipulated models of normal tissue. One may even learn enough about cancer to be able to control it. This last and very unscientific thought, so foreign to most pure scientists, is foremost in the mind of anyone who has cared for terminally ill patients suffering from cancer.

In writing an outline for the book, I decided to take a new approach. An analogy may be useful. There are those who are interested in fish scales. They want to know about the development, function, and chemical composition of scales. Can you imagine an individual studying fish scales if he had never even thought about a fish or had possibly never even seen one? It is in this position that most authors of books about cancer have placed nonmedical people. The authors have completely ignored the clinical phenomenology of cancer to discuss a highly specialized aspect of it. There are many scales, but no fish. The first chapter of this book was to be a short description of the clinical behavior of tumors. My intent was not to make diagnosticians of the graduate students, but to provide breadth and allow for more intelligent scientific probing. This was to be followed by discussions of the developmental aspects of neoplasms with incursions into carcinogenesis, control processes, metastases, and immunology, all of which were to be reviewed in light of developmental concepts.

After outlining the strategy, I began reading and writing with vigor. It will surprise no one to learn that I was soon overwhelmed by the volume and complexity of information to be evaluated and incorporated in the book. Fortunately, Louis Fink, expert in chemical carcinogenesis and molecular biology of neoplasms, and Bob Shikes, expert in developmental pathology, agreed to coauthor the volume. Together, we decided what to put in and what to omit, resisting pressure to respond to the "current hot trends" in favor of developing a theme. Information was then evaluated according to the theme. There have been many arguments about whether or not we were being fair and attained a reasonable balance. We decided to include a bibliography so that students wishing more specialized information could acquire a foothold in the literature. This raised another perplexing and never fully resolved problem about which experiments to note. In a short volume, it was obvious that not all meritorious information could be quoted. So finally, we made no attempt to be inclusive. Rather, we have referenced only those articles that contribute directly to the theme of development and neoplasia.

There will be those who believe that there is too much preoccupation with the developmental approach and the concepts that tumors are caricatures of the process of normal tissue renewal and carcinogenesis is a model of differentiation. Since the concepts make sense, are supported by large amounts of data, and open vistas for meaningful new experiments, the text will stand. The developmental approach allows one to find unity in apparently diverse facts and stresses the fundamental commonalities peculiar to all tumors. Finally, a short chapter on therapy is included to give insight into the state of that art. Brief descriptions of commonly studied tumors are added, hopefully, to make it easier to understand reports that take for granted that the students know what a "hepatoma" really is.

We are grateful to three of our students, Sara Clarke, Linda McManus, and Chris Robinson, for reading the manuscript and making helpful comments. Our associates, John Lehman, Paul Nakane, Harlan Firminger, and Hank Fennell, all made inputs of great value. We thank them. Anna Williams, LaVonne King, and Vicky Starbuck typed the manuscript and the countless revisions. Without their understanding this task would have been a nightmare.

We revised and revised and corrected and corrected and finally came to the conclusion that we could work on this small volume forever. So one day we sent it to the publisher, and now it is yours.

Denver, Colorado   G. BARRY PIERCE

ONE

Introduction

Cancer is responsible for more than 300,000 deaths per year in the United States, and during our lifetime more than 50 million people will be treated for this disease. Since the disease is usually chronic and often fatal, and it affects people of any age, the cost in both economic and human terms is enormous. Thus, it is not surprising that a vast research industry has been geared to the study of cancer: identifying and eliminating causes, elucidating mechanisms, and defining exploitable differences between normal and neoplastic tissues. This industry has produced a mass of information staggering in its scope and volume. Tumors have been studied as problems of metabolism (specifically in terms of nucleic acids, proteins, carbohydrates, and enzymes) and as problems of virology, genetics, endocrinology, and virtually every other discipline of biology. The result is an informational overload of such proportions that it has become difficult to integrate advances even in a single discipline, much less in the entire field of oncology.

This problem is compounded by a variety of other considerations. There are many kinds of cancer and remarkable differences in the incidence of cancer among species, subspecies, and strains. Within the human species there are striking variations determined by sex, age, race, and geography (Tables 1–1 and 1–2). Tumors may occur in almost any site in the body and in these varied locations give rise to complex clinical manifestations. There are multiple kinds of tumors, multiple causes for tumors, and multiple diseases caused by tumors. These diversities are so great that some think of neoplasms as disparate entities with few commonalities. The summation of these effects has been to endow cancer with a mystique. As more and more information is accu-

1

TABLE 1–1

AGE-ADJUSTED DEATH RATES FROM CANCER/100,000 POPULATION

	All Cancer Sites		Breast Cancer	Colon Cancer	Stomach Cancer	Prostate Cancer
	Men	Women				
U.S.A.	150	100	20	20	10	15
Austria	190	120	20	25	35	15
Japan	140	90	5	10	60	2
Venezuela	110	100	10	5	35	10

SOURCE: Based on data from Seidman, H., E. Silverberg, and A. I. Holleb, Cancer Statistics, *CA—A Cancer Journal for Clinicians*, 26 (1976), p. 2. Numbers have been adjusted to nearest multiple of 5.

TABLE 1–2

INCIDENCE OF CANCER BY ANATOMIC SITE IN U.S.A.

Common	Uncommon	Rare
Lung (bronchus)	Brain	Heart
Colon	Bone	Small intestine
Breast	Kidney	Adrenal gland
Uterus	Testis	Parathyroid

mulated through our disease-oriented approach, it becomes increasingly difficult to form a unifying concept of cancer.

The concept of cancer used here is based on the long-known fact that cancer is an abnormal tissue composed of cells derived from a tissue of the host. Since all other tissues arise from pre-existing cells by the process of cell division and differentiation, it would seem logical to suppose that a cancer develops by a process paralleling the development of normal tissues. When the fragmentary and disease-oriented knowledge of cancer is considered from this developmental perspective, order is discovered among apparently unrelated facts. In other words, when viewed from a developmental standpoint, cancer begins to make sense.

The strategy of this chapter will be to describe cancer in the traditional terms of its clinical manifestations, biological behavior, and affect on the host. This will serve as a basis for the ensuing discussion of cancer as a problem of developmental biology.

Pathology of neoplasms

The term *tumor* dates from medical antiquity and denotes a swelling or lump. In this sense a common boil is a tumor, but the tumors with which we are concerned are new growths or *neoplasms*, which are composed of parenchymal cells and a supportive stroma of blood vessels and connective tissues. Neoplastic parenchymal cells may develop from any cell of the body capable of mitosis. The cells may multiply slowly, with the neoplasm remaining small, or rapidly and progressively, forming a mass of such proportions that the host literally cannot maintain it. Whether fast or slow, the rate of growth of the cells exceeds that of adjacent tissue. However, the notion that tumor cells are the fastest growing cells of the body is not true. The rate of normal cell renewal in the testis, intestine, and bone marrow is directly comparable to the rate of cell division in most rapidly growing tumors.

Since neoplastic cells originate from normal cells, it is not surprising that they should bear some structural and functional resemblance to the cells and tissues of origin. Those that bear closest resemblance to the normal tissue usually form slowly growing, encapsulated masses described as *well differentiated*. The cells of an islet cell tumor of the pancreas may closely resemble normal islands of Langerhans' cells and synthesize insulin. A tumor arising in cartilage may resemble cartilage and synthesize chondromucoprotein and a tumor of the colon may resemble colonic mucosa and synthesize mucin. Whereas the normal tissue performs these highly specialized functions under controls, meeting the physiologic needs of the host, neoplastic cells may not respond to usual physiologic regulation. For example, an islet cell tumor may synthesize excessive amounts of insulin causing coma and death, or one of the bowel may produce pools of unneeded mucin.

Although differentiated features of the tissue of origin may be carefully reproduced in neoplastic parenchyma, it is more usual for differentiated characteristics to be diminished or absent. Thus, a tumor of islet cells of the pancreas may not synthesize insulin, one of the colon may neither form glands nor synthesize mucin, or one of muscle may bear scant resemblance to normal muscle and its cells may synthesize little myosin. Such neoplasms are usually rapidly growing and are described as *undifferentiated*; they lack normal morphologic characteristics and specialized cellular function.

Neoplasms have the capacity to transgress upon adjacent normal tissues. A slowly growing, well differentiated tumor grows expansively and may destroy adjacent tissues by compression or physical displacement, or cause obstruction of blood vessels and ducts, or erode into blood vessels causing hemorrhage. In each of these situations, physical signs and symptoms of disease will be produced, depending upon the anatomical locale of the mass. Some of the cells of undifferentiated tumors have the additional capacity to invade and destroy adjacent normal tissues. For example, undifferentiated tumor cells in the brain might compress, invade, and destroy a vital center, leading to death. A small focus of undifferentiated tumor cells in the mucosa of the bowel has the capacity to invade the muscle layers and eventually grow around and constrict the lumen of the bowel causing intestinal obstruction (Figure 1–1). In a similar manner, invading neoplasms may obstruct bronchi, bile ducts, or any other ductal system. Neoplastic cells may invade blood vessels, lymphatics, and body cavities and be disseminated

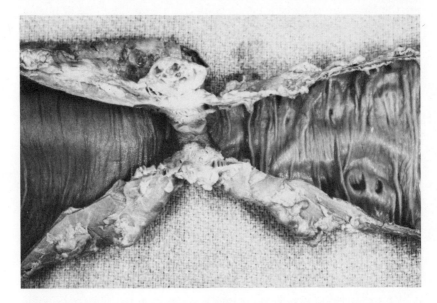

(a)

(b)

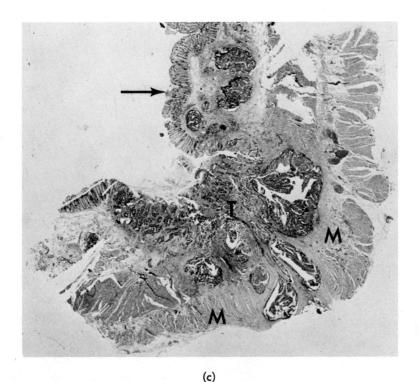

(c)

Fig. 1-1 Adenocarcinoma of the colon. (a) The bowel has been cut longitudinally to reveal a concentric ring of tumor tissue that markedly narrows the lumen. (b) Sections through the intestinal wall show gray streaks of tumor originating in the mucosa (arrow) and invading the wall to its serosal surface. (c) Microscopic section to show normal mucosa (arrow) and tumor (T) that has invaded from a focus in the mucosa through the muscle (M) to the serosal surface.

to distant parts of the body where they implant and grow as secondary tumors (Figure 1-2). These secondary growths invade and destroy the tissues at the site of implantation. This property of tumors—disseminated, discontinuous growth—is known as *metastasis* and is the most feared attribute of neoplastic cells.

All neoplasms have the capacity for continual growth, usurping the nutritional resources of the host even at times of dire need by the host. Lipomas (benign tumors of fatty tissue) have been observed to grow progressively in inmates of concentration camps who were being starved to death. Not only were the tumors able to monopolize the resources of the host, but the fat stored in them was unavailable as a source of nourishment. Thus, the tumor usurps from the host and contributes nothing to it in a regulated manner. This constant drain upon the indi-

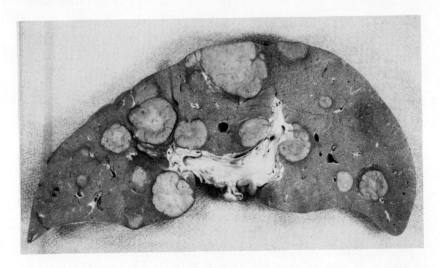

(a)

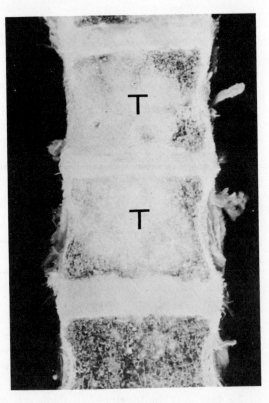

Fig. 1–2 (a) Metastases in liver. Although the liver is among the most frequent sites for metastasis, every organ in the body is susceptible to this process. (b) Metastases to vertebral bone. The presenting symptomatology is sometimes due not to the primary tumor but to its metastases. This patient came to his physician because of back pain (due to the destruction of vertebral bone) and anemia [due to replacement of bone marrow by gray masses of tumor (T)]. The vertebra at the bottom has only a few small foci of gray tumor. The primary tumor was subsequently found to be a small carcinoma of the prostate.

(b)

vidual's resources may be adequately compensated for if the tumor is small or slowly growing, but if the tumor is large or rapidly growing, *cachexia* (a condition of starvation) results. The mechanism of cachexia is unknown. However, the net effect is that resistance is lowered and the host perishes from infection, usually pneumonia or septicemia. Approximately one-half of terminal cancer patients die of infection. There is evidence that some neoplasms may synthesize toxic materials that poison the host and contribute to the cachexia found terminally. These materials, called *toxohormones*, have been studied by Nakahara and Fukuoka (1948), but they have not been completely characterized.

Certain tumors have synthetic capabilities not found in the tissue of origin. Tumors of nonhormone-producing tissues may synthesize hormones; rarely, a tumor of the lung may synthesize ACTH (the adrenocorticotrophic hormone normally synthesized in the pituitary gland that stimulates the adrenal gland to synthesize cortisone) or HCG (human chorionic gonadotrophin normally synthesized by trophoblast). Mesothelial tumors may synthesize an insulin-like factor (Hall, 1974). Excessive production of such potent hormones results in easily recognized clinical symptoms. Tumors probably synthesize other factors that do not result in clinical manifestations and such factors are, therefore, usually undetected. In this regard, antigens have been identified in tumors of the bowel originally thought to be specific for those tumors (Gold and Freedman, 1965). Whatever their specificity, they have been called *carcinoembryonic antigens* because they are synthesized by embryonic and neoplastic cells. The significance of these phenomena will be discussed later (see p. 66).

Tumors are clinically dangerous for two general reasons. The positional-functional danger is that an otherwise innocuous, slowly growing, well differentiated tumor may develop in a vital structure or synthesize a biologically potent material in quantity, leading to the demise of the host. The other danger is in the intrinsic attributes of the tumor cells, which result in a mass of rapidly proliferating, undifferentiated cells capable of invasion and causing cachexia and death.

Classification and nomenclature

The most widely used classification of neoplasms is based upon the effect of the tumor upon the host. Tumors that normally do not destroy their host or are, at worst, capable of causing the destruction of the host from a purely positional-functional standpoint are known as *benign tumors*. *Malignant tumors* consist of cells that are intrinsically dangerous because they are rapidly growing and quickly cause cachexia

' invade and metastasize. The intrinsically dangerous cells of
ınt tumors invariably destroy their host unless they are either
extirpated or killed.

This behavioral classification is usually superimposed upon an
anatomical one, in which the tumor is identified by the tissue of origin.
The suffix *-oma* when appended to the name of a tissue means a benign
tumor of the particular tissue; for example, a *fibroma* is a benign tumor
of fibrous tissue, a *chondroma* is a benign tumor of cartilage, and an
adenoma is a benign tumor of any glandular epithelium. Malignant
tumors are divided into two broad categories on the basis of the tissue
of origin. To denote a malignant tumor of epithelial origin, the root
carcino- (meaning crab-like) is combined with the suffix *-oma*. Thus, an
adenocarcinoma of the stomach is a malignant tumor of the glands of
the stomach. Similarly, to denote a malignant tumor of mesenchymal
origin, the root *sarc-* (meaning fleshy) is used with the suffix *-oma*. Thus,
a *fibrosarcoma* is a malignant tumor of fibrous tissue. In lay terminology all malignant tumors are called *cancer*, a reflection of the infil-
trative growth that reminded our forefathers of crab-like tentacles
(Table 1–3).

TABLE 1–3

NOMENCLATURE

	Benign			*Malignant*	
		Example			*Example*
Epithelial	—polyp	of colon		—carcinoma	squamous cell carcinoma of skin
	—adenoma	of liver			adenocarcinoma of colon
					undifferentiated carcinoma of lung
Non-epithelial	tissue + oma	leiomyoma		tissue + sarcoma	leiomyosarcoma
		lipoma			liposarcoma
		osteoma			osteosarcoma
Blastoma	—	—		tissue + blastoma	nephroblastoma
					neuroblastoma
Teratoma	—teratoma	of ovary		—teratocarcinoma	of testis

The principal reason for distinguishing carcinomas from sarcomas is that the carcinoma has a tendency to spread first by the lymphatics, whereas sarcoma spreads by the vasculature. This distinction is of clinical significance only.

There are a few exceptions to this terminology. *Blastoma* is an archaic term that refers to highly malignant tumors that histologically resemble embryonic tissues. Thus, a *neuroblastoma* is a highly malignant tumor consisting of cells resembling neuroblasts. *Melanoma* is a malignant tumor of pigment-synthesizing cells and properly should be termed *melanocarcinoma*. Similarly, *hepatoma* is commonly used to denote *hepatocarcinoma*, and *myeloma*, a malignant tumor of plasma cells, denotes *myelosarcoma*. Leukemia, which literally means "white blood," is a malignant growth of leukocytes characterized by the circulation of such large numbers of neoplastic leukocytes in the peripheral circulation that the blood appears creamy in color (see Appendix, p.185).

Benign and malignant tumors

Many attempts have been made to develop methods for unequivocal diagnosis of malignancy with accurate prediction of the patient's outcome or prognosis. The best of these, but by no means perfect, are gross and microscopic examination. The previous experience of the pathologist has prepared him to know the biological behavior to be expected from the various gross and microscopic patterns of neoplastic tissue.

Gross examination of neoplasms is often less revealing than microscopic examination, but invasion or metastasis is characteristic of malignancy. Sarcomas usually have a brain-like consistency with areas of hemorrhagic necrosis, whereas carcinomas are usually extremely hard and may contain areas of calcification. Malignant tumors may be encapsulated, but these are usually pseudocapsules penetrated by invading tumor cells. One of the most useful aspects of gross examination is to *stage* the degree of development of the tumor. If an adenocarcinoma is grossly confined to the organ stage (stage 1), the prognosis is significantly better than if it has invaded adjacent tissue (stage 2). Stage 3 tumors have metastasized to some of the regional lymph nodes; stage 4 tumors are those with distant visceral metastasis and, with rare exceptions, these individuals are doomed.

Benign tumors have a fibrous capsule or grow out from a surface and never show evidence of invasion or metastasis. From a histologic standpoint, the cells of benign tumors closely resemble the normal differentiated cells of the particular organ [Figs. 1–3 (a) and 1–4 (a)]. Evidences

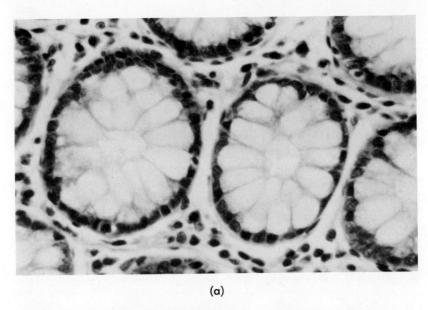

(a)

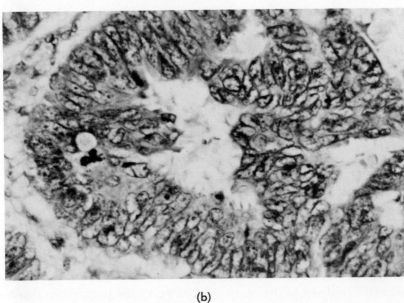

(b)

Fig. 1-3   (a) This tumor of the colon is composed of glands that resemble those of normal colon. Specialized function is evidenced by large quantities of mucin within the cells. Such close correspondence to non-neoplastic colon allows the pathologist to designate the tumor as benign. Contrast this with the tumor shown in (b). Although a gland-like structure is suggested, the cells are piled up, crowded, dyspolaric, and pleomorphic. There is little evidence of specialized function and several abnormal mitoses are present. These features (anaplasia) denote the likelihood of invasion and metastasis. Such a tumor is designated as malignant.

of rapid growth (mitotic figures, multiple large nucleoli, indented pleomorphic * nuclei, amphophilic † cytoplasm) are absent [Figs 1–3 (b) and 1–4 (b)]. The well differentiated cells may be organized into tissues closely resembling the normal organ with no microscopic evidence of invasion or metastasis.

The microscopic characteristics used by the pathologist to make a diagnosis of malignancy are collectively known as *anaplasia,* a term literally meaning "without form" [Figs. 1–3 (b) and 1–4 (b)]. An anaplastic or undifferentiated malignant tumor is made up of cells that lack many of the overt features of the original differentiation. For example, the cells of an adenocarcinoma may be arranged in poorly shaped glands; if more anaplastic, they may not be arranged in glands, but in cords and columns. In extreme situations the cells, although still capable of synthesizing mucin, may grow singly, apparently lacking the cohesiveness of epithelial cells. Nuclear cytoplasmic ratios are high and the nuclei are usually abnormal in shape with multiple, densely staining nucleoli. The cytoplasm is amphophilic and there are numerous and often bizarre mitotic figures. Microscopic evidence of invasion of blood vessels, normal tissues, or lymphatics by tumor cells are all unequivocal signs that these cells are intrinsically dangerous. Death of the host will result unless these cells are removed or killed.

In arriving at a prognosis, the pathologist must consider many factors, including the staging of the tumor and the degree of anaplasia. Each case must be considered individually, with the prognosis representing the sum of a variety of pluses and minuses. Thus, a patient who has a highly anaplastic tumor that has not yet spread and can be removed in its entirety has an excellent prognosis. A patient who has a well differentiated tumor that cannot be removed in its entirety or has already metastasized by the time of diagnosis has a poor prognosis. These major considerations may be modified by other factors, such as the patient's age, that have been empirically determined by answering the question, "How have similar tumors behaved in similar hosts?"

The implication from the above is that in theory there is no single criterion for malignancy, a fact that has been known to pathologists for generations but is not apparent to many oncologists who search for the *sine qua non* of malignancy. Instead of relying upon a single criterion for the diagnosis of malignancy—for example, number of mitoses, nucleocytoplasmic ratio, lack of differentiation, or staining characteristics—pathologists use a summation of characteristics which, when found together in a particular tumor, indicate a diagnosis of

---

* Having multiple shapes and forms.
† Stainable with acid or basic dyes.

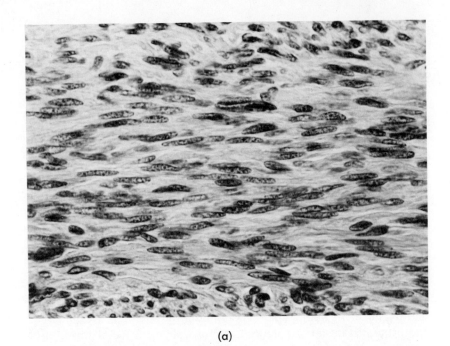

(a)

(b)

Fig. 1–4 Here comparison is made between two non-epithelial tumors, both derived from smooth muscle. (a) The resemblance to non-neoplastic smooth muscle is apparent. (b) Although there is some resemblance to smooth muscle, the features of anaplasia are prominent. Benign behavior is expected of the highly differentiated tumor and malignant behavior of the anaplastic tumor.

malignancy. From a practical standpoint, evidence of invasion or metastasis in an adult animal is diagnostic of malignancy.

Clinical examples

The following are typical case histories to illustrate some of the biological problems of oncology.

A 35-year-old woman consulted her physician because of a lump in her breast. The lump was freely movable and rubbery on palpation, and there was no involvement of regional lymph nodes. An extirpational biopsy was performed and the pathologist found the tumor to be encapsulated and made up of cells microscopically resembling normal breast tissue. The diagnosis was adenoma of the breast, and the patient was cured by removal of the mass.

Contrast this history with that of a 48-year-old woman with a painless, nontender, hard lump in the breast. The mass measured 3 centimeters in diameter and seemed fixed to the underlying muscle and overlying skin, which suggested that the cells of the tumor had invaded these structures. Two small, hard nodules were found by palpation in the axilla (one of the routes of lymphatic drainage of the breast). At surgery the pathologist found an infiltrating, unencapsulated mass; microscopic analysis revealed that it was composed of anaplastic cells. The pathologist found metastases in four regional lymph nodes removed from the axilla. About 50% of people with this type of history survive for about 5 years and are considered cured. The others develop widespread metastases, particularly to the lungs, liver, ovaries, adrenal glands, and bone, and they usually die of a terminal infection, often a pneumonia. This particular patient remained well for 14 months and then began to lose weight rapidly. She became weak and metastatic tumors were identified radiologically in her lungs and bones. She was treated with x-rays and chemicals, but other metastases developed and she became emaciated and died. At the autopsy of this cachectic woman, widespread metastases were found in the lungs, liver, adrenals, and vertebral column, but the immediate cause of death was pneumonia. The tumor had exhausted her resources and rendered her unable to combat a simple infection.

TWO

The Early Development
of Neoplasms

Introduction

Most human cancers are discovered only when they have grown to a recognizable gross size, or invaded, or metastasized; in other words, when they cause disease. Thus, with the possible exception of superficial tissues such as skin or cervix, it is impossible to observe the early events in human neoplasms. What is known about early development of tumors has been obtained from studies using animals, with the hope that these observations will extrapolate to the situation in man. However, there are differences in behavior between tumors of rodents and tumors of man. Most rodent tumors do not invade or metastasize. They are encapsulated, but their cells are as anaplastic as those of metastasizing tumors of man, and on the basis of morphologic characteristics, growth rate, and ability to kill the host by cachexia, they are considered to be malignant. As pointed out in Chapter One, the phenomenon of malignancy is impossible to define succinctly. Some of the tumors that have been considered malignant in rodents may be benign. In a short-lived species, many of the manifestations of malignancy as they are known in man may not appear before the animal succumbs to cachectic effects.

It is of some interest that when metastases from extremely malignant human tumors are heterotransplanted to cortisone-treated hamsters, irradiated rats, and the anterior chamber of the guinea pig's eye, they do not metastasize. They grow as encapsulated masses but with all of the features of anaplasia characteristic of the primary tumor. Possibly there is something about the local environment in the host that prevents some of these heterotransplants from metastasizing.

The problem of whether or not rodent tumors must invade and metastasize to be considered malignant may be insignificant in view of the discussion of the origin of benign and malignant tumors in Chapter Five.

Development of tumors in animals

Initiation, latency, and promotion

Yamagiwa and Ichikawa (1918) confirmed experimentally the clinical observations made by Potts in 1795 that soots and tars were related to the high incidence of skin cancers of chimney sweeps. Prolonged application of tars to the ears of rabbits produced warty papillomas and, eventually, metastasizing carcinomas. The first tumors did not appear for 15 months, and it is now known that, irrespective of the chemical carcinogen or the age of the animal, either a prolonged period of time, or a latent period, always precedes the development of tumors.

Rous and Kidd (1941) discovered that the early development of a tumor was not necessarily a uniformly progressive process; instead, it was characterized by a series of steps that ultimately resulted in progressively growing and metastasizing cancer. Painting the skin of a rabbit's ear with carcinogen resulted first in the appearance of benign, warty tumors. With continued painting, clinical cancer might develop. If painting were discontinued, the lesions might regress. When regression was complete, further painting with carcinogen again elicited tumors at the exact site of the previous ones. Such a cycle of development and regression of tumors could be repeated, but eventually a mass of neoplastic cells developed that grew progressively without remission until the host died of cachexia, septicemia, or metastasis.

The discontinuous growth observed early in chemically induced tumors is also evident in virally induced tumors. Although newborn mice receive the mammary tumor virus when they first suckle the mother, there is a latent period of approximately 40 weeks before tumors appear. Some tumors grow rapidly during pregnancy but regress or even disappear between pregnancies. This cycle of growth and regression may continue, but eventually a noncyclic progressive pattern of growth becomes established that may begin during the regressive periods or during the periods of rapid growth.

The characteristics of the long latent period and the intermittent growth that occurs before establishment of the unrelenting, continuous growth typical of established neoplasms have been the subject of intensive study and some of the contributing factors have been elucidated. For example, carcinogens may vary in their ability to produce tumors. Some are so potent that they will cause clinical cancer after a single application to the skin of a rodent. Others, which are described as being weak, require repeated doses of the agent. Friedwald and Rous (1950) discovered that when a weak carcinogen is applied to the skin, a heritable change occurs in a number of cells that may persist indefinitely. These altered cells do not develop into clinical cancer and persist unnoticed in the skin. If this skin is repeatedly treated with the weak carcinogen, or if it is nonspecifically stimulated by wounding or by treatment with irritants, a neoplasm will develop. This nonspecific stimulation affects the altered cells directly because if skin is treated once with weak carcinogen and is grafted to a second animal and then wounded, a tumor will develop in the skin. By this type of experimentation, it was discovered that this first change caused by carcinogen occurred quickly, persisted indefinitely, and was heritable from one generation of cells to another. It was called *initiation* by Friedwald and Rous (1950).

Berenblum and Shubik (1949) investigated the interaction of carcinogenic and noncarcinogenic irritants in the formation of tumors and concluded that carcinogenesis is a two-step process. The first is initiation of latent tumor cells; the second, promotion by agents that in themselves are not carcinogenic. The promoter commonly used was Croton oil (an irritant extracted from the seeds of *Croton tiglium*, a leafy shrub), the active ingredients of which are phorbol * esters. The most notable one is 12-0-tetradecanoyl-phorbol-13-acetate (TPA), which in concentrations of 10^{-9} molar per application to epidermis caused basal cells (the stem cells of the epidermis) to divide, resulting in epidermal hyperplasia. The promoter is capable of stimulating both carcinogen-initiated and normal stem cells to divide. Thus, it has been hypothesized that tumor production is caused by the proliferation of stem cells by selective gene activation that is associated with marked changes in plasma membrane-associated functions. The extremely small amounts of TPA necessary for promotion suggest a hormone-like activity with the possibility of specific receptor sites on plasma membranes. The fate of TPA within the cell is unclear (Sivak, Mossman, and Van Durren, 1972).

Promoting agents frequently have both lipophilic and hydrophilic

---

* Phorbin is the porphyrin nucleus of chlorophyll.

activities. Some investigators believe that they interact with the lipid of the cell membrane (Estensen *et al.*, 1974). Radioactive TPA binds to membranes and causes a stimulation of the membrane enzyme, Na+, K+-dependent ATPase, and an increase in membrane phospholipid synthesis. Several other events occur in cells after TPA treatment. These include increased and new RNA synthesis, increased protein synthesis, increased and altered histone synthesis, stimulation of DNA synthesis and cell division, a decrease in cyclic AMP, and a rapid increase in cyclic GMP. Since this change in cyclic GMP has been associated with mitogenic stimulation, it is possible that it triggers a multitude of secondary events in the cell (Belman and Troll, 1974; Boutwell, 1974).

Both steroids and antiproteases appear to be capable of blocking the promoting effects of the phorbol esters (Troll, Klassen, and Janoff, 1970). This is of interest because within 1 to 3 hours after TPA treatment of cells in culture, plasminogen activator is found to increase (Wigler and Weinstein, 1976). The role of proteases in cell division is discussed in Chapter Four and may be relevant in the mechanism of action of promoters.

In addition to wounding, there are many agents capable of causing promotion, including Tween, anthralin, iodoacetic acid, fatty acid esters, phorbols, turpentine, hormones, and viruses. For that matter, it is difficult at times to know what is initiating and what is promoting. For example, Duran-Reynals (1952) showed that when subthreshold doses of carcinogen are painted on the skin of rodents, tumors will develop at the site of painting if nonspecific, non-oncogenic viruses are administered to the animals. He painted methylcholanthrene on the skin of rats, followed this with exposure to pox virus, and obtained tumors in these animals. Viruses are usually considered initiators; can some of them function as promoters?

Raick (1974) has shown that normal skin treated with phorbol derivatives quickly assumes an appearance comparable to squamous cell carcinoma. Although epithelial growth is rapid, this "promoted-but-uninitiated skin" reverts to normal with cessation of treatment and does not form cancer, at least not under the conditions of the experiments. There are important implications in these observations for interpretation of what has been considered to be precancerous lesions.

From such observations it may be concluded that although promotion may be difficult to define, it does not, in the absence of initiation, cause cancer. The principal action of promoters seems to be the stimulation of cell division resulting in a shortened latent period, although Raick (1974) has evidence that more than stimulation of cell division occurs in promotion.

Repeated application of carcinogen increases the number of tumors and shortens the latent period. Dose response curves can be constructed for each carcinogen—increased numbers of tumors are produced by increased amounts of carcinogens. The increase in the number of tumors produced suggest that more latent neoplastic cells are produced which, by promoting effects of the weak carcinogen, yield more tumors.

Critical mass and environment factors in dormancy and latency

Experiments with promoters have led us to the realization that environment controls neoplastic expression after initiation. Although little is known about the control of latent malignant cells, the observation that they do not immediately form a tumor may have a parallel in embryonic development. Grobstein and Zwilling (1953) have shown that it takes a critical number of cells of like potential to form a tissue. They studied the cellular dynamics of head formation in the chick embryo. After 25 hours of incubation, the cells of the head anlagen, removed from the embryo and placed in tissue culture, would proliferate, differentiate, and form a head. If the anlagen were divided into quarters or eighths and each fragment were cultivated individually, a head would form from each fragment; but if smaller fragments were cultivated, masses of undifferentiated nonspecific cells without differentiation into a head would result. If the small fragments were pushed back together and cultivated, a head would develop. Thus, it appears that sufficient cells of like nature create an environment uniquely suitable for their particular needs and optimal function as a tissue. In this case, it is known that successful growth of cells *in vitro* depends upon a relatively large explant with a suitable ratio of explant to fluid medium. Similarly, a critical number of cells must be inoculated for successful transplantation of a tumor. The corollary to this idea is that subthreshold numbers of cells may respond negatively to their environment and be incapable of expressing their phenotype. This is probably the situation with initiated cells. A long latent period is required for them to attain a critical mass and express the malignant phenotype. Many bits of information support this concept. As mentioned, it has been shown in dose-response experiments that small amounts of carcinogen produce a few tumors with long latent periods, whereas larger doses of the same carcinogen produce more tumors with shorter latent periods. Large or repeated doses of carcinogen either affect more cells (by initiation of more cells or by simultaneous promoting action), thereby achieving a threshold number, or continue the intracellular changes of initiation resulting in an autonomous cell capable of proliferating rapidly and forming a tumor. The latter idea is less compatible with the facts of dormancy than the idea that a threshold number of like cells is required to produce a tissue.

Latency has a close parallel in the clinical phenomenon of dormancy, which is best illustrated in rare cases of adenocarcinoma of the human breast. Occasionally, individuals treated for adenocarcinoma of the breast by radical surgery appear cured and free of disease for 10 or more years. Then a tiny nodule, 2 or 3 millimeters in diameter, develops in the surgical scar or in a remote metastatic site. Biopsy confirms the original diagnosis. Moreover, the new tumor is identical to the original one. Irrespective of the type of therapy, the course is usually rapid, terminating in death due to metastasis and cachexia.

Since the breast had been removed 10 years earlier and there is no breast tissue in a scar, the best explanation of dormancy is that a subthreshold number of adenocarcinoma cells remained in the mastectomy wound and were entrapped in the scar. This environment was inhospitable and the cells were unable to express their malignant phenotype. Constrained by their environment, the cells remained dormant and required years to produce enough malignant cells to create an environment optimal for the expression of the malignant phenotype, *i.e.*, critical mass.

Many elderly men have foci of latent cancer cells in their prostate glands (Moore, 1935) that are morphologically identical to the cells found in other patients who have metastatic adenocarcinoma of the prostate. The incidence of both the quiescent foci and metastasizing adenocarcinoma correlate directly with age; yet these cells in the one situation remain dormant for years and do not behave as malignant cells.

A useful model of dormancy has been developed by Fisher and Fisher (1959, 1967a) who related the development of liver metastasis in the rat to the injection of known numbers of cancer cells into the portal vein. Two hundred fifty thousand malignant cells always resulted in metastasis and death in 2 or 3 weeks, but when as few as 50 of these cells were injected in the portal circulation, the animals survived indefinitely. When the livers of these survivors were examined by laparotomy, no evidence of tumor was noted. Yet, many of them developed liver metastasis 2 or 3 weeks after surgical exploration. Apparently some of the 50 tumor cells remained dormant in the liver and were stimulated to divide and produce tumors by the surgical trauma and handling of the viscera. In addition to wounding and other nonspecific trauma, Fisher and Fisher (1967b) found that certain chemicals and hormones were capable of promoting the growth of dormant cells.

Thus, it appears that tumor cells placed in a hostile environment may respond negatively in terms of growth rate and remain so until environmental conditions change or until the neoplastic mass becomes large enough to provide its own optimal growth conditions. The phenomenon of dormancy involves a few fully developed tumor cells that have a code of behavior inflicted on them by the environment, and it is probable that latent tumor cells in the prostate are also bona fide cancer cells responding to their environment. When the environment is changed by chemical irritants (weak carcinogens, promoters, surgical trauma), the environment of the cell is changed to one favoring proliferation. The net result is the attainment of a threshold mass that creates its own microenvironment, resulting in expression of the malignant phenotype. Once this occurs, the system locks in, and it is difficult for the therapist to overcome this situation.

The size of the critical mass varies widely among tumors. Since some highly malignant tumors can be *cloned* (a new tumor produced from a single cell); obviously, under those conditions, the threshold number is one.

Latency has important practical and theoretical implications. Studies of chemical carcinogenesis have always been hampered by the latent period necessary for the development of tumors. This makes it virtually impossible to study the "process" of carcinogenesis at a molecular level. Carcinogenesis is an event or series of events related in time, and if the molecular nature of the process is to be understood, it must be studied in a situation in which the time factor can be managed (see Transformation, Chapter Six).

Progression, dependency, autonomy

Whereas Rous and his associates emphasized the discontinuous nature of neoplastic development, Greene (1957) stressed the notion that tumors progress through a series of stages that eventuate in an autonomous state. To Ewing (1940), autonomy meant a lack of response to controls, and this led to the erroneous notion that tumors were "lawless" tissues. To Greene, autonomy meant that tumor cells were no longer dependent upon the original environment and could survive and proliferate in foreign environments. Greene (1957) discovered that embryonic tissues and highly malignant tumors could be successfully transplanted to the immunologically protected environment of the anterior chamber of the eye of a heterologous species. Survival in this situation represented, presumably, the ultimate in the autonomous state. Greene studied the development of spontaneous uterine and mammary adenocarcinomas of rabbits; he discovered that the early lesions grow slowly, do not metastasize, and do not heterotransplant to the anterior chamber of the guinea pig's eye. Later the growth rate of the tumor increases, the tumor metastasizes, and concomitant with the latter change, the tumor acquires the ability to survive after heterotransplantation. Thus, Greene postulated that malignant tumors continue to gain neoplastic attributes and become progressively independent of the regulating mechanisms of the host. When they can metastasize or survive in a heterologous host, they are considered to be autonomous.

Huggins demonstrated that androgens are required for maintenance and normal function of the prostate gland. Elderly dogs, like elderly men, develop benign cystic hyperplasia of the prostate, and Huggins reasoned that anti-androgen therapy should change the environment of the cells and reduce the hyperplasia. This is exactly what happened, and in search of better therapeutic approaches to adenocarcinoma of the prostate, he castrated patients and administered estrogens (Huggins and Hodges, 1941). In many cases there was dramatic clinical improvement. The metastases, which characteristically affect the spine causing intractable pain and destruction of bone, regressed. Men near death recovered and returned to work. The usual lifespan of a man with carcinoma of the prostate with metastasis is often from only 9 to 16 months, although there is considerable variation. Rarely some individuals survive 10 years or more. In Huggins' first series of 20 cases, 4 survived for 12 years. However, despite such dramatic regressions, there were many failures. This therapy was not curative; the adenocarcinomas eventually recurred and the patients died. For a period of time the cells of these neoplasms were dependent upon the presence of testosterone for their continued growth. Removal of testosterone plus the inhibitory action of estrogen were responsible for creating an environment in which many of the tumor cells could not thrive. Ultimately this response was lost, but it is not known if the loss of responsiveness resulted from mutation with production of unresponsive cells or if cells unresponsive to estrogen were present from the inception of the tumor and eventually overgrew the responsive cells in the system. Huggins was the first to modulate growth of tumors by altering cellular environment.

Subsequently, Foulds (1956, 1969) studied spontaneous and transplantable mammary carcinomas of mice and confirmed Rous' concept of conditional neoplasms. As described previously, some spontaneously developing mammary adenocarcinomas would grow rapidly during pregnancy, disappear between pregnancies, and recur with subsequent pregnancy. Eventually, these tumors developed a continual and progressive growth pattern that terminated in the death of the host. Certain mammary adenocarcinomas grew well when transplanted to suitable females, but they did not grow in males. Ovariectomy inhibited growth of these tumors, but castration of male mice did not abolish their refractoriness unless estrogens were also administered. The dependency upon estrogen was not absolute in any of the tumors, since many implants in normal mice began to grow after long dormancy. This was particularly true if large inocula were used in transplanting the tumors. In the intervals between pregnancies some of the regressed tumors recurred spontaneously without evident cause and then grew independent of pregnancy.

The lessons learned from the work of Huggins, Rous, and Foulds have been applied to the treatment of carcinoma of the breast in attempts to find tumors responsive to antagonistic endocrine therapy. Only about 20% of treated individuals show any sign of therapeutic remission with estrogen therapy and many of these remissions are minimal. Some tumors appear to lack receptor sites for hormones. For example, in certain breast tumors the lack of response to therapy with estrogen appears to be correlated with the loss of binding capacity for this hormone.

It is important to realize that endocrinologically responsive tumors are true neoplasms, not a nonspecific hyperplasia of normal cells resulting from endocrine stimulation. The experiments of Furth (1953, 1967–68) have clarified this matter. Furth administered the goitrogen, thiourea, which, as shown by MacKenzie and MacKenzie (1943), inhibits the synthesis of thyroxin, thus eliciting by a feedback mechanism an increased secretion of thyroid-stimulating hormone (TSH) by the pituitary gland. This in turn produces a marked hyperplasia of the nonfunctioning thyroid. As Furth correctly reasoned, adenocarcinomas eventually developed in the hyperplastic glands. Some of these adenocarcinomas metastasized to the lungs and resulted in death. Continued growth of either the primary tumor or the metastasis was dependent upon high levels of TSH. If TSH levels were depressed by the administration of thyroid hormone, the tumors regressed and even disappeared, only to reappear when TSH levels were elevated again. The tumors that were dependent upon high levels of TSH would not transplant into normal animals, but would grow in an animal that had been previously conditioned by thyroidectomy. The latter caused increased levels of TSH. After repeated transplantations of these *dependent* tumors into conditioned hosts, the tumors lost their dependency upon TSH and grew progressively in normal animals.

Biskind and Biskind (1945) transplanted fragments of testis into the spleens of castrated rats. Interstitial cell tumors, dependent upon high levels of pituitary gonadotropin, developed in the transplants. The mechanism depends upon interference with the feedback control that testosterone normally exerts on gonadotropin synthesis by the pituitary gland. Castration of the animal removed testosterone, and the testosterone produced by the graft was carried by the portal circulation to the liver where it was metabolized before it could inhibit gonadotropin production by the pituitary gland. In the absence of the usual circulating levels of testosterone, the pituitary gland synthesized increased levels of gonadotropin which stimulated the graft in the spleen and resulted in the development of *dependent* interstitial cell tumors.

The endocrine system is the easiest one in which to observe the phenomenon of dependency, but many other kinds of dependency exist. One of the notable ones is the dependency of some tumors upon asparagine (Broome, 1961). Whereas normal cells can synthesize enough asparagine to meet their requirements, a few tumor cells cannot and they depend upon their hosts for adequate amounts of this amino acid. In the presence of asparaginase (the enzyme that hydrolyzes asparagine to aspartic acid plus ammonia), the neoplastic cells are deprived of asparagine and die, but normal cells survive.

A similar explanation may account for the high incidence of lymphoreticular tumors in immunosuppressed people bearing organ transplants. One could envision inhibition of a feedback mechanism akin to that of the endocrines to account for their production. In the absence of a good immune response (due to steroid or cytotoxic therapy to prevent graft rejection), the foreign antigen continually bombards the primitive cells of the lymphoreticular system to produce immunocytes. Immunocytes proliferate, but they cannot produce antibody or react against the foreign cells because of the therapy. A reticulum cell sarcoma results. The development of these tumors has been considered to be the result of the lack of immune surveillance as a result of immunosuppression, but it probably has nothing to do with surveillance at all (see Chapter Nine).

There are probably other examples of *interdependence* of tissues for maximal function because there must be integrating mechanisms that allow for normal development of organs. Since these are unknown, their role in carcinogenesis is also unknown.

A common denominator is present in all of the studies described—tumors gain or lose characteristics with time. Based on his long-term studies of breast cancer, Foulds (1969) proposed the concept of *progression*. This concept states that tumors gain or lose unit characteristics independently, and these changes are irreversible and result in autonomy. Unit characters include ability to metastasize, growth rate, ability to synthesize molecules, and characteristics of differentiation. Two possibilities could explain the phenomenon of progression: (1) It could be the result of overgrowth of existing cells best able to survive under the conditions that would impart their characteristics on the tumor (selection of preexisting cells) or (2) Heritable changes might be induced in cells that would then overgrow the populations.

The ascites has been useful for studying progression and autonomy (see Chapter Ten). Conversion of a solid tumor to the ascites, with repeated intraperitoneal passage of free-growing tumor cells, results in remarkable changes in the tumor line. Usually there is an increase in growth rate, an increase in ability to metastasize, and a loss of differentiated functions. Since these attributes occur independently of each other and since they are irreversible changes, they meet the definitional requirements of progression. Klein and Klein (1956) and others have provided evidence that these changes result from a selection of preexisting cells best able to survive under the conditions of the ascites.

Gray and Pierce (1964) examined the relationships of growth rate, differentiation, and progression in a malignant melanoma of the hamster. This tumor, first described by H. S. N. Greene, was maintained through many transplant generations, and it almost invariably metastasized to the lungs. Although most cells were pigmented, occasional metastases were amelanotic, grew rapidly, and never regained the ability to synthesize melanin. Support for the notion that this tumor was made up of heterogeneous collections of cells with different capacities for proliferation and differentiation was obtained by cloning. Six clonal lines were obtained. One was slowly growing and densely pigmented and never metastasized. Others had intermediate growth rates and degrees of pigmentation corresponding to those of the primary tumor. Another grew rapidly, metastasized widely, and was amelanotic. Amelanotic tumors never regained the ability to form pigment, and when they invaded a blood vessel and metastasized, the secondary tumors were amelanotic. Identification of these lines with their differing characteristics all derived from the same tumors was strong support for the concept that selection played an important role in ascites conversion and in progression. Those tumor cells best able to survive under the highly selective conditions of the intraperitoneal environment would eventually outgrow the others and their characteristics would dominate in the resultant tumor. It would be interesting to know why, in the primary tumor lines, cells with such different properties were able to coexist and grow as tumors without rapid segregation. Obviously, cells influence each other, either by direct cell-cell contact or by producing an environment that provides control, as in dormancy.

As mentioned, investigators studying chemical carcinogenesis have postulated that repeated doses of carcinogen or promoting agent may complete the changes of initiation within cells. There is little information on this point. One of the features of neoplasms that has received too little attention is the instability of the neoplastic karyotype compared to that of normal cells. Possibly the degrees of ploidy and chromosomal rearrangement continually going on in a tumor may contribute to the genetically heterogeneous population with evolution of autonomous and malignant cells, which then overgrow the others.

To determine whether or not changes that could explain progression could be induced in cells *in vivo*, we attempted ascites conversion of the deeply pigmented, slowly growing melanoma cloned from the melanoma of Greene (Gray and Pierce, 1964). Minces of tumor were injected intraperitoneally and failed to grow. Then cells were injected intraperitoneally, left for 24 hours, washed from the cavity, and injected subcutaneously. The resultant tumor had been selected for survival in the intraperitoneum and the process was repeated many times. Eventually a rapidly growing amelanotic melanoma was obtained. This sequence of events suggests that prolonged exposure to the hostile environment of the peritoneal cavity induced new properties in melanoma cells, which were then selected for by repeated intraperitoneal passage. Selection of a mutant is also an explanation for these observations. However, it must be noted that mutation is not the only means by which heritable changes may occur in cells. Sonneborn's experiments with paramecia (1959) illustrate an alternative possibility.

When paramecia of suitable genetic type contain a certain cytoplasmic particle (termed *kappa*), they secrete an antibiotic that is lethal for paramecia lacking this particle. Kappa appears to be a degenerate parasite that replicates independently of the cell. When paramecia are grown under conditions of rapid proliferation, colonies devoid of kappa can be obtained. Not only are these *undifferentiated* animals unable to produce the antibiotic, they are also sensitive to its harmful effects. If kappa is reintroduced into their cytoplasm through mating, the animals once again synthesize the antibiotic and are resistant to its effects. This raises the analogous possibilities that tumors are selected for cells best able to proliferate under the conditions of their environment and they may divide faster than their cytoplasmic components can be synthesized. The tumor would thus gradually lose cytoplasmic features characteristic of the original differentiation and would be unable to reinitiate them because the genome would be locked into other channels. The loss of cytoplasmic features characteristic of the original differentiation would be irreversible. This loss of differentiated attributes should not be construed as a dedifferentiation because the cells have not gained potential (see Chapter Five).

A note of caution must be maintained when interpreting these data. The response of cells under extremely stressful situations has been shown. Whether or not this reflects the mechanism of autonomy *in vivo* is unknown.

In summary, tumors usually appear long after the initiating stimulus, an interval termed *latency*. During the latent period a threshold or critical number of malignant cells is produced and creates an environment conducive to the optimal growth of tumor cells. During latency the initiated cells respond negatively to the host environment, unless, of course, the latter changes in a manner that is hospitable to tumor cell growth. The latent interval can be shortened by promoting substances that stimulate malignant cells to divide by changing this cellular environment. When the neoplastic cell mass reaches threshold numbers, the cells no longer remain dormant but grow rapidly and express the malignant phenotype.

The conclusion is inescapable that tumor cells are not uncontrolled, but respond positively or negatively to environmental factors—most of which are unknown. Thus, *autonomy* is a misnomer.

During the life of the neoplasm, characteristics are gained or lost through a process called *progression*. Progression has two components: (1) a situation in which individual cells lose or gain characteristics, thereby contributing to the heterogeneity of the cellular composition of the tumor, and (2) the selection of a particular subpopulation of cells best able to proliferate under the conditions. The characteristics of this group of cells will eventually replace those of the original tumor.

Certain neoplasms are conditional in the sense that they are dependent upon environmental, growth-promoting substances. With continued transplantation in suitable hosts or with time in the same host, dependence upon the trophic hormone or growth promoting substance is lost. This *autonomy* in turn may be construed as a manifestation of progression. This would be favored by Fould's concept. The concept of stimulation of dormant cells to form a tumor stresses the controlling influence of environment on neoplastic expression. The neoplastic phenotype would be expressed when enough neoplastic cells aggregate to create an optimal microenvironment. The development of a tumor is probably the result of a progression of cellular changes plus environmental control. Some believe that initiated cells have a gene mutation and promoters cause these cells to complete the expression of the mutation. Others believe that promoters act through cell division. Obviously, much more information is needed, particularly in the area of the effects of the environment upon cells.

THREE

Tumors As Caricatures
of Tissue Renewal

Introduction

In considering neoplasms as outlined in Chapter One, the only concept that evolved was that malignant neoplastic tissue is made up of cells derived from tissues of the body. Some features of the original differentiation may be present, others may be absent, and new ones may be added. Tumors may cause disease by positional-functional effects secondary to the presence of the mass; however, of greater clinical importance is the fact that the cells of malignant tumors have attributes that are dangerous and ultimately lead to the demise of the host. Since one of the notable features of an established neoplasm is progressive growth uncoordinated with that of adjacent normal tissue, neoplasms have been considered aberrations of growth. Because malignant tumors are usually faster growing and less differentiated than benign ones, the notion that a neoplasm represents an aberration of growth is coupled to another idea—the neoplasm is also a problem of differentiation.

The dogma of the inevitability of death from untreated malignancies is one that must be accepted. Spontaneous remission is rare and a clinical curiosity; patients who anticipate such a miracle are doomed. The observation that neoplastic cells always give rise to more neoplastic cells has naturally led to the notion that "once a cancer cell, always a cancer cell." Few have considered the possibility that malignant parenchyma may contain other than malignant cells.

Four prototype tumors will be discussed in this chapter. These tumors contain multiple stem lines, each with different properties that play an important role in determining what the final phenotype of the established tumor will be. It will be seen that malignant stem cells do have a capacity for differentiation; in some cases, benign cells that correspond to their non-neoplastic counterparts evolve. Thus, the dogma "once a cancer cell, always a cancer cell" is false. Tumors are made up of a heterogeneous mixture of undifferentiated, partly differentiated, and fully differentiated cells. This leads to a working concept of neoplasms—a tumor is a caricature of its tissue of origin in both appearance and mode of development.

Differentiation in tumors

Teratocarcinomas
of mammals

The term *teratocarcinoma* literally means "malignant tumor resembling a monster or malformed baby." These tumors usually occur in the gonads (rarely in other sites) and are exceedingly malignant—by the time a mass is first noticed the tumor has often metastasized, most often to the periaortic lymph nodes, liver, and lungs (Dixon and Moore, 1953). These tumors are so lethal that survival rates are measured by 2 years rather than the usual 5 years. By then the host is either dead or cured.

Although many believe that malignant stem cells have some ability to differentiate, proof of this contention was lacking and the literature is replete with the obverse idea—dedifferentiation. The first direct evidence for differentiation of malignant cells was obtained in studies of testicular teratocarcinoma. These tumors are uniquely suited to these studies because they are made up of multiple somatic tissues (Fig. 3–1). Many of the cells are well differentiated and arranged in easily recognized tissues which may form primitive organs, *e.g.*, spinal cord, gut, skin, or teeth. The most extreme examples of this type of organization resemble embryos and have been termed *embryoid bodies* (Peyron, 1939). The embryoid bodies of murine teratocarcinomas resemble

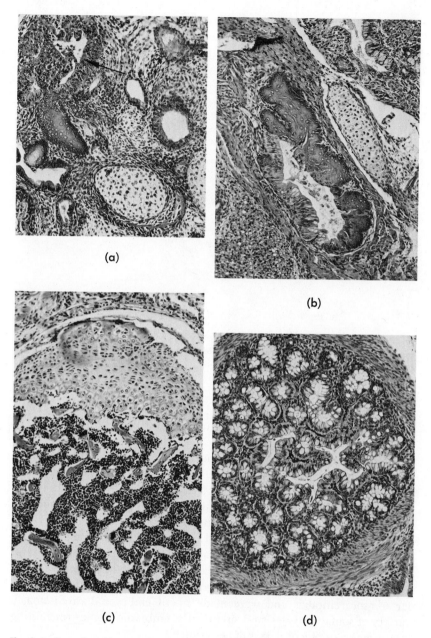

(a)

(b)

(c)

(d)

Fig. 3–1 Transplantable testicular teratocarcinomas from strain 129 mice, illustrating degrees of differentiation and organization in the tumors (92×). (a) Note the undifferentiated embryonal carcinoma (arrow), cartilage bone, squamous epithelium (skin) glands, and muscle cell in disorganized array. (b) Note the cartilage, squamous, and glandular epithelium, again disorganized (92×). (c) Note the organization: There is a cartilagenous cap, endochondral ossification, bone, and marrow. (d) Note the glands surrounded by muscle. This small "organ" was found in masses of neural tissue. From Pierce, Dixon, and Verney (1959), courtesy of Lippincott Co.

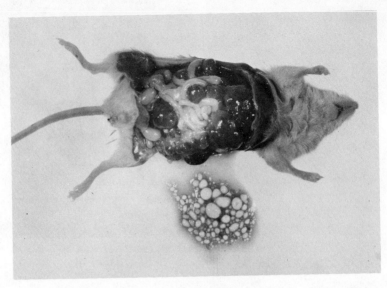

Fig. 3–2   Strain 129 mouse, injected intraperitoneally four weeks previously with teratocarcinoma. Note the large vesicular embryoid bodies below the animal. From Pierce (1961),
courtesy of Academic Press, Inc.

mouse embryos (Pierce and Dixon, 1959; Stevens, 1959) (Fig. 3–2) and
those of human teratocarcinomas, human embryos, lending validity to
the notion that they are embryoid.

Randomly interspersed among the somatic tissues of the teratocarcinoma are embryonal carcinoma cells (highly malignant cells) that resemble embryonic epithelium (Fig. 3–1). If the embryonal carcinoma
cells could differentiate into the other tissues of the teratocarcinomas,
the transition should be easily documented.

This documentation was obtained by a combination of embryologic
methods and by cloning transplantable testicular teratocarcinomas of
strain 129 mice, originally developed by Stevens and Little (1954).
These tumors appeared to be the exact counterpart of human testicular teratocarcinomas and were made up of embryonal carcinoma, a
variety of somatic tissues, and, frequently, embryoid bodies resembling
mouse embryos of 5 or 6 days of gestation.

Attempts to separate embryonal carcinoma from the somatic tissues
resulted in a method for the mass production of embryoid bodies
(Pierce and Dixon, 1959). Since only highly malignant cells will form
an ascites (see Ascites Tumors, p. 169), conversion to the ascites was attempted as a means of selecting for the embryonal carcinoma. The

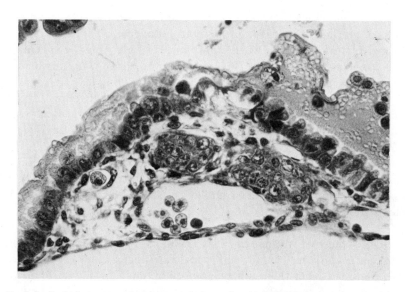

Fig 3–3   Typical structure of a large vesicular embryoid body (320✕). Representatives of all three germ layers are present. Note the investing layer of endoderm, the mesenchyme (with sinuses containing hematopoietic cells), and embryonal carcinoma (Pierce, Dixon, and Verney (1960), courtesy of Paul B. Hoeber Inc. © 1960, U.S.–Canadian Div.–International Academy of Pathology.

teratocarcinoma did convert, but the embryonal carcinoma did not proliferate as single cells in the peritoneal fluid; instead, the fluid abounded in small aggregates of cells (Fig. 3–2). These proved to be embryoid bodies (Pierce and Dixon, 1959) (Figs. 3–3 and 3–4), the most

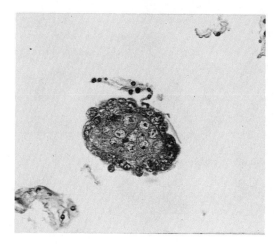

Fig. 3–4   An immature embryoid body made up of a central mass of embryonal carcinoma overlain by endoderm (300✕). From Kleinsmith and Pierce (1964), courtesy of Cancer Research Inc.

immature of which were made up of solid cores of embryonal carcinoma overlain by a layer of visceral endoderm or yolk sac. They formed teratocarcinomas when transplanted singly into the subcutaneous space of strain 129 mice (Pierce and Dixon, 1959). Thus, embryoid bodies made up of embryonal carcinoma and endoderm were capable of forming a teratocarcinoma made up of a wide collection of epithelial, endodermal, and mesenchymal derivatives. Similar evidence was obtained by Stevens in developmental studies of teratocarcinomas (Stevens, 1959).

More mature embryoid bodies were large and cystic and, in addition to endoderm and embryonal carcinoma, contained mesenchyme. The mode of development of these complex embryoid bodies was studied in the ascites (Pierce and Dixon, 1959) and *in vitro* (Pierce and Verney, 1961). Formation of mesenchyme within them was found to resemble embryonic mesenchymogenesis (Figs. 3–5 and 3–6). In other words, when proximal endoderm overlaid embryonal carcinoma, a few of the latter cells detached from the main mass of embryonal carcinoma and differentiated into mesenchyme. This mesenchyme was capable of differentiating into vascular sinuses containing primitive hematopoietic cells. Other mesenchyme formed cartilage or bone.

Very large, cystic embryoid bodies often lacked embryonal carcinoma. In a statistical study it was found that the incidence of teratocarcinomas resulting from transplantation of these embryoid bodies correlated directly with the incidence of embryonal carcinoma in the embryoid bodies (Pierce, Dixon, and Verney, 1960). The development of teratocarcinomas from cystic embryoid bodies was studied at daily intervals after transplanting single embryoid bodies subcutaneously. It was

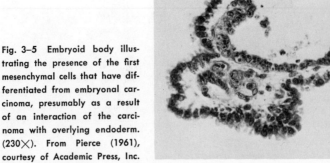

Fig. 3–5 Embryoid body illustrating the presence of the first mesenchymal cells that have differentiated from embryonal carcinoma, presumably as a result of an interaction of the carcinoma with overlying endoderm. (230✕). From Pierce (1961), courtesy of Academic Press, Inc.

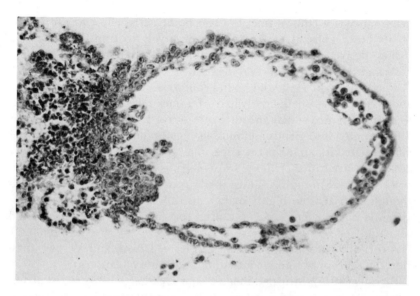

Fig. 3–6 An embryoid body grown from an explant of a teratocarcinoma *in vitro* (230✕). Note the investing layer of yolk sac, the underlying mesenchyme with sinusoids containing hematopoietic cells, and embryonal carcinoma (230✕). From Pierce (1961), courtesy of Academic Press, Inc.

found that the endoderm and mesenchyme of the embryoid body had the capacity to differentiate, but they did not form a tumor. The embryonal carcinoma cells, however, proliferated rapidly, invaded through the endoderm of the embryoid body and into the subcutaneous tissue of the host, and formed a mass. Differentiation into the multiplicity of tissues characteristic of the tumor subsequently took place in this mass of embryonal carcinoma cells. Thus, it was concluded that embryonal carcinoma cells were the multipotential stem cells of teratocarcinomas (Pierce, Dixon, and Verney, 1960).

Confirmation for this idea was obtained by cloning single embryonal carcinoma cells (Kleinsmith and Pierce, 1964). For these experiments, small embryoid bodies of embryonal carcinoma and endoderm were dissociated with trypsin and single cells were transplanted into the peritoneal cavity of mice. Eleven percent of the single cells grew into teratocarcinomas, each containing from 12 to 15 somatic tissues. Since endoderm was incapable of forming a teratocarcinoma, by exclusion the embryonal carcinoma cells were the multipotential stem cells of teratocarcinomas.

It was important to determine whether or not the tissues derived from embryonal carcinoma were benign or malignant. The concept of

malignancy is a clinical one; the end point is the effect of the tumor
on its host. Thus, it was necessary selectively to transplant somatic
tissues, derived from but free of embryonal carcinoma, into suitable
hosts and measure the effect on the host.

The large cystic embryoid bodies containing endoderm mesenchyme,
and occasional epithelial derivatives, but lacking embryonal carcinoma,
were transplanted subcutaneously into strain 129 mice. The endoderm
differentiated into glands and mucous membrane, and the mesenchyme
differentiated into muscle, cartilage, and bone. These well differentiated
tissues survived in their hosts for as long as 6 months; during this interval they did not grow. Thus, it was concluded that these tissues, although derived from highly malignant cancer cells, were benign. Apparently the dogma "once a cancer cell, always a cancer cell" did
not hold for teratocarcinomas (Pierce, Dixon, and Verney, 1960). It
was this observation that first made Pierce, Dixon, and Verney question the concept of somatic mutation as a cause of cancer.

Brinster (1975), Mintz, Illmensee, and Gearhart (1975), and Papaioannou, McBurney, and Gardner (1975) have extended the observation
that some of the progeny of malignant stem cells differentiate into benign postmitotic senescent cells. Each has introduced a small number
of embryonal carcinoma cells into a blastocyst from a different strain
of mouse. The grafted blastocysts were returned to the uterus and
allophenic * mice bearing the morphologic and biochemical characteristics of both the tumor and recipient strain were born. One of the
allophenic mice was a male that sired more than 40 offspring bearing
the genetic traits of the teratoma line. Thus, it can now be concluded
that progeny of malignant stem cells can be normal.

Teratocarcinomas of plants

It has been established that certain bacteria, viruses, and chemicals can cause the development of redundant masses of rapidly proliferating disorganized cells in certain plants (White and Braun, 1942).
These are called *galls* and appear to be the equivalent of malignant
neoplasms in higher forms of life.

In addition to these unorganized growths, others have been described
that appear analogous to teratomas. Plant teratomas have abnormal
leaves, buds, and occasionally rootlets that appear at the margins of
disorganized masses of rapidly growing cells. Not surprisingly, there
was controversy over the origin of the differentiated structures of these
tumors. Were the leaves and buds the result of differentiation and or-

* An allophenic mouse is an artificially produced mosaic comprised of at
least two genetically different ancestral lines.

ganization of neoplastic cells, or were they derived from normal cells carried along in the growth of the tumors?

Years of experimentation were required to resolve this question. Final proof of the contention that the differentiated structures were an integral part of the tumor and derived by differentiation from the neoplastic cells was obtained by cloning teratoma cells *in vitro*. Single cells grew into typical teratomas with disorganized cells, leaflets, and buds. The differentiated structures not only looked normal, further examination proved them to be normal (Braun, 1969).

When teratomatous buds were grafted to the rapidly growing tips of normal plants, galls were produced that bore leaves. When these galls were repeatedly grafted to other rapidly growing normal shoots, they eventually formed shoots which grew into plants when placed on special media. The plants matured, and flowered, and some bore fertile seed. The tumor that gave rise to plants bearing fertile seeds usually had a normal number of chromosomes; the infertile one had abnormal numbers. Irrespective of ploidy, cells and tissues derived from teratoma cells were capable of differentiation and organization into flowering plants.

In studies of the metabolic requirements of plant tumor cells in tissue culture, Braun (1956, 1969, 1972) discovered that the major difference between neoplastic and normal cells lies in the ability of the neoplastic cells to synthesize at least two essential growth factors (auxin and kinetin). These are present in embryonic plant tissues, but with differentiation their production is repressed. In the tumor, the genes responsible for the production of these factors are apparently derepressed and lead to a level of proliferation that leads to the development of a mass.

This is then another demonstration that the neoplastic change is a caricature of normal tissue genesis with proliferation outweighing the differentiative and organizational capacity of the component cells. The development of normal cellular offspring from neoplastic cells again argues against the dogma "once a cancer cell, always a cancer cell."

Squamous cell carcinoma

As a further test of the concept that the progeny of stem cells of a cancer have a capacity for differentiation, a squamous cell carcinoma * of the lip of a rat was studied (Pierce and Wallace, 1971). This tumor, induced chemically, was transplantable, grew slowly, did not metastasize, and consisted of undifferentiated cancer cells with

---

* See Appendix, p. 193; for clinical data.

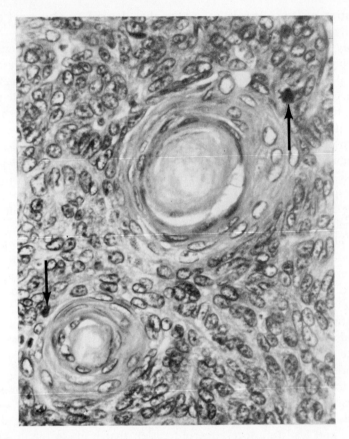

Fig. 3–7 A squamous cell carcinoma of the rat, illustrating the appearance of two squamous pearls and masses of undifferentiated carcinoma (600×). From Pierce and Wallace (1971), courtesy of Cancer Research Inc.

many mitoses, and squamous pearls * (Fig. 3–7). The pearls were made up of easily recognized squamous cells lacking mitoses.

DNA of the stem cells was labeled by injecting tumor-bearing animals with tritium-labeled thymidine, and tumors were harvested at intervals to follow the fate of the labeled cells (Figs. 3–8, 3–9, and 3–10). At 2 hours after injection, only undifferentiated cells outside the pearls were labeled; those in the pearls were not. After 50 and 96 hours, more and more labeled cells appeared in the pearls. Thus, the undif-

* When a bit of normal squamous epithelium is transplanted subcutaneously, it forms a nodule closely resembling the squamous pearl; thus, the neoplastic pearl may be considered the equivalent of normal differentiation and organization into squamous epithelium.

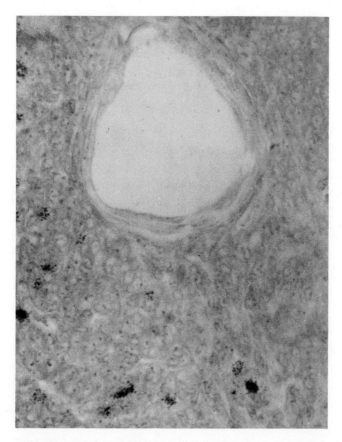

Fig. 3–8   Radioautogram 2 hours after administration of thymidine-^{3}H. The squamous pearl is not labeled; numerous undifferentiated cells are labeled (600✕). From Pierce and Wallace (1971), courtesy of Cancer Research Inc.

ferentiated cancer cells were believed to have migrated into the pearls, confirming the observations of Frankfurt (1967). Since labeled cells in the pearls contained fewer grains than those in the undifferentiated cells, the labeled stem cells must have divided at least once before differentiation and incorporation into pearls.

Since undifferentiated cancer cells are capable of invasion, it was conceivable that some of the labeled stem cells might have invaded pre-existing pearls. To explore this possibility, the experiments were repeated and the state of differentiation of the cytoplasm of labeled cells was determined by using autoradiography with the electron microscope. Cells labeled at 2 hours had exceedingly undifferentiated cytoplasm characterized by a lack of membranous organelles with large

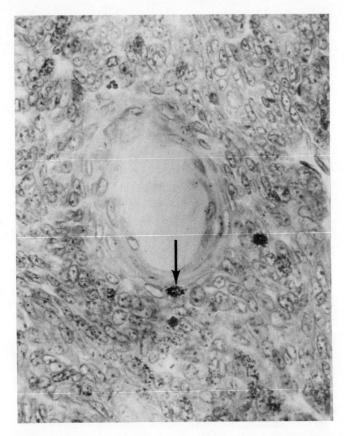

Fig. 3–9   Radioautogram of a labeled nucleus (arrow) in a pearl 50 hours after the administration of a pulse of thymidine-^{3}H (600$\times$). From Pierce and Wallace (1971), courtesy of Cancer Research Inc.

numbers of free ribosomes and polysomes. In addition, the plasma membranes lacked the differentiated features characteristic of squamous cells. This appearance was in contrast to that of labeled cells in the pearls that resembled normal squamous cells.

The tumor-forming capacity of undifferentiated tumor cells and cells of squamous pearls was compared by transplanting equal amounts of these tissues into appropriate hosts. Seventy-eight single pearls were transplanted subcutaneously and a like amount of the undifferentiated squamous cell carcinoma was transplanted in other animals. None of the pearls developed into a tumor, whereas 27 of 82 undifferentiated transplants formed typical squamous cell carcinomas. It was concluded that the undifferentiated stem cells of this tumor differentiated

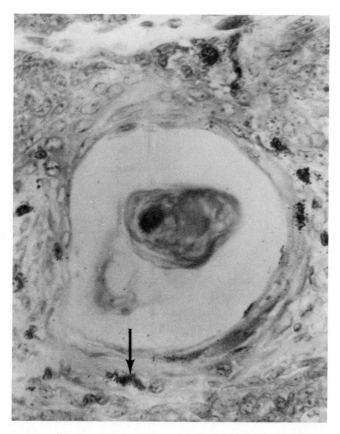

Fig. 3–10 Radioautogram of labeled nuclei (arrows) in a pearl 50 hours after the administration of a pulse of thymidine-^{3}H (600$\times$). From Pierce and Wallace (1971), courtesy of Cancer Research Inc.

into squamous cells, which were incapable of synthesizing DNA and incapable of forming a tumor. They were benign (Pierce and Wallace, 1971). Just as the teratocarcinoma was a caricature of embryogenesis, the squamous cell carcinoma was a caricature of tissue genesis.

Neuroblastoma

Neuroblastoma (see p.187 for clinical details) is a highly malignant tumor that afflicts children. It is composed of cells resembling the neuroblasts of early development. Embryonic neuroblasts differentiate into ganglion cells, and the possibility that neuroblastoma cells might differentiate into ganglioneuroma cells was an intriguing possibility.

Equally interesting was the rare occurrence of the spontaneous disappearance of incurable neuroblastoma and the marked difference in prognosis in very young patients in relation to older ones.

Goldstein *et al.* (1964) explanted human neuroblastomas *in vitro.* Initially, explanted cells were round and lacked processes, but once attached to a substrate, they sprouted axons and dendrites, particularly when closely packed. By electron microscopy, these were well differentiated cells, resembling ganglion cells, with processes that ended in terminal boutons. Unlike most cells in tissue culture that round up when dividing, these highly complex cells with entangled processes were capable of cell division in their complex, spread-out form. The cells synthesized and degraded catecholamines. Dopa was converted into dopamine and norepinephrine, and the metabolites of the amines were released into the medium (Goldstein, 1972). The cells were stimulated by nerve growth factor and contained neuronal proteins specific to the central nervous system.

Yokoyama *et al.* (1971) studied human neuroblastomas ultrastructurally and observed that although the primary tumors were made up of rounded cells, they had many fine cytoplasmic processes. Large tumor cells contained prominent rough endoplasmic reticulum, numerous free ribosomes, microfilaments, and microtubules. The cytoplasmic processes contained these organelles plus large coated vesicles approximately 1,000 angstrom units in diameter and small coated vesicles 500 angstrom units in diameter. These vesicles are typical of nerve processes. Schwann cells and myelinated nerve fibers were not present, although some cells had infolding of the plasma membranes. Synapse-like structures have been described in neuroblastomas (Misugi, Misugi, and Newton, 1968). One of the tumors was chromaffin positive and one had large quantities of dopamine.

A wide range of differentiation was found when a ganglioneuroblastoma was examined ultrastructurally. Typical ganglion cells and Schwann cells were found in addition to poorly differentiated neuroblastoma cells.

Extensive experiments have been carried out on a neuroblastoma of the mouse, C-1300, isolated at the Jackson Laboratory. This is an undifferentiated tumor made up of small round cells with little overt differentiation (rosette formation). Goldstein (1972) has demonstrated that when these rounded cells were attached to glass, they differentiated, as evidenced by the extension of nerve processes and by the enlargement of nucleus and cell bodies. Axonation was stimulated by the addition of BudR, cyclic AMP, and prostaglandin E_1 to the medium. Addition of 10% serum had the tendency to suppress neurite outgrowth, as did explantation to collagen substrates (Miller and Levine, 1972). They also noted no decrease in growth rate in cells that had undergone axonation in contradistinction to the observation of Seeds *et al.* (1970). In this respect there is little or no difference in tumor-forming capacity between cultures of neuroblastoma cells with cell processes and those grown in circumstances favoring the rounded-up conditions. Thus, it is conceivable that the degree of axonation observed in tissue culture is a modulatory rather than a differentiative effect (p. 71). Be that as it may, the observation that cells of a highly undifferentiated tumor make a phenotypic change *in vitro* is an important contribution that merits intensive study.

Leukemia

Since leukemia * is characterized by the presence of malignant blast forms as well as well differentiated elements, one could postulate that it was a caricature of the development of leukocytes. The mode of development and maturation of blood-forming cells had been controversial. One theory of their origin would have each of the leukocytic and erythropoietic tissues derived independently. Each separate precursor cell, through cell division and maturation, would supply the requisite number of differentiated cells in response to environmental signals. The other idea, for which there is a good deal of support, is that there is a common, overall stem cell for all of the blood-forming elements. Evidence for the existence of the pluripotential stem cell compartment comes *in vivo* cloning experiments using the spleen assay technique developed by Till and McCulloch (1961). The pluripotent stem cell gives rise to a group of precursor stem cells, each committed for a particular differentiation. Thus, the precursor compartment for the myelopoietic series is derived from the pluripotent hematopoietic stem cell compartment. Within the myelopoietic series, differentiation occurs into each type of mature nondividing myelocyte. This compartment responds to environmental perturbation, but it is restricted in its capacity for self-renewal and it depends upon input from the pluripotential stem cells for its maintenance. Erythropoietic tissues presumably also differentiate from pluripotential stem cells and are regulated through the action of erythropoietin.

The existence of the myeloid compartment as a nonselfsustaining compartment was discovered through studies of marrow cells and leukemias *in vitro*. Bradley and Metcalf (1966) and Ichikawa, Pluznik, and Sachs (1966) individually developed methods for growing murine hematopoietic cells in agar. In these experiments, feeder cells (spleen and kidney cells that synthesize a colony-stimulating factor, CSF) were

* See Appendix, p.185 for a clinical discussion of leukemia.

suspended in a semisolid tissue culture medium containing enough agar to make a stiff gel. Marrow cells were suspended in a soft agar overlay containing media. The cultures were incubated for approximately 14 days. Under these circumstances, single cells proliferated and, if a colony of 50 or more was produced, the cells were designated as having originated from a colony-forming cell (CFC). If fewer cells were produced, they were referred to as *clusters*, not colonies.

These methods were adapted for cultures of human marrow by Pike and Robinson (1970). Cloning of leukemic cells in this system gave rise to colonies containing granulocytes and macrophages, confirming the notion that these cell types were capable of differentiating from common precursor cells as indicated in the spleen colony assay of Till and McCulloch (1961). When 160 sequential granulocytic colonies, each originating from a single cell, were classified by Metcalf and Moore (1970) on the basis of their component cells, 6% of the colonies consisted of blast cells; 60% consisted of myeloblasts and myelocytes with no evidence of metamyelocytes or polymorphs; 30% were mixed colonies with cells ranging from myeloblasts to polymorphs, and 6% were colonies consisting solely of polymorphs. Again, the multipotency of a single precursor cell was established.

Whang *et al.* (1963) used the Philadelphia chromosome (see p. 121) as a marker for the identification of human leukemic pluripotential stem cells, which appeared to be capable of erythropoietic, granulopoietic, and megakaryocytic differentiation. When the disease was in relapse and progressing rapidly, differentiation into granulocytic pathways was increased relative to that for erythropoiesis. When the disease was in remission, there was less granulopoiesis and more erythropoiesis.

These results were confirmed by Ichikawa (1970) who found that in the presence of conditioned media, phagocytosis was induced in the progeny of myeloid leukemic cells. This evidence of differentiation correlated with a loss of leukemogenic potential. He concluded that two factors were present in the cultures: a heat stable factor required for growth and a heat labile factor responsible for differentiation.

Paran *et al.* (1970) observed that acute myeloid leukemic cells (undifferentiated "blast" cells) grown in semisolid agar media underwent differentiation not normally seen *in vivo*. CFC were 350 times more numerous in peripheral blood of leukemics than in that of normal people and gave rise to huge colonies containing differentiated myeloid elements. Thus, the *in vitro* environment was conducive to differentiation apparently not possible *in vivo*. Greenberg, Nichols, and Schrier (1971) grew cells from patients with acute myeloid leukemia and obtained innumerable small colonies. Ultimately these colonies consisted exclusively of mature granulocytes. Since the donor marrow was originally made up of malignant blast cells, it was concluded that they had in fact differentiated.

Moore *et al.* (1972) believe that there is a block in cell differentiation *in vivo* in leukemia. In acute myelocytic leukemia, defective differentiation was observed during relapse, but normal differentiation returned during remission. It was postulated that the leukemic clone was suppressed during remission and, indeed, normal colony-forming cells were detected in cultures. In tissue cultures derived from patients with chronic myelomonocytic leukemia, polymorphs were present early and were subsequently replaced by large numbers of macrophages—a pattern of differentiation similar to that seen in normal marrow. The observation confirmed the notion that granulocytic and monocytic cells have a common ancestry. Finally, these experiments demonstrated the capacity for differentiation of leukemic cells, a process analogous to the situation in solid tumors.

The colony-stimulating factor (CSF) is important in the *in vitro* growth of hematopoietic cells. This glycoprotein is present in many tissues of the mouse and its highest concentration is in the submaxillary gland. Chervenick and LoBuglio (1972) have obtained evidence that this material is made by monocytes. CSF is an essential requirement for establishment of colonies of marrow cells *in vitro* and for continuous growth of these cultures. Although it appeared initially that human marrow cells might be an exception to this requirement, Metcalf, Bradley, and Robinson (1967) showed that, intermingled with the marrow cells, were cells capable of synthesizing CSF. If these cells were separated from the myelopoietic elements, the myelopoietic elements became dependent upon the addition of CSF to the culture medium.

Granulocytic colonies give rise to macrophages in the presence of CSF inhibitors. When large amounts of CSF are present, there is stimulation of growth of myeloid as well as macrophagic elements. This might suggest that CSF is, in fact, a growth factor that may secondarily result in a degree of suppression of differentiation. CSF is active at extremely low concentrations, in the range of action of hormones. Its action is specific in the sense that CSF derived from human urine will stimulate marrow but not other cells. CSF levels are high in the serum of leukocyte-depleted subjects. Low levels of activity are found in leukemic patients, but this may be due to the presence of inhibitors.

A particularly interesting tumor, erythroleukemia, is produced by the injection of Friend leukemia virus into mice. The tumor disseminates widely and lines of undifferentiated cells have been isolated that show little or no differentiation. If these cells are injected into lethally irradiated mice, they colonize the spleen, and undergo differentiation.

produce hemoglobin, and they are benign. However, other cells do not differentiate and retain their malignant properties (Friend, Preisler, and Scher, 1974). Cell lines have been derived from erythroleukemia cells (Scher, Preisler, and Friend, 1973) and in the presence of dimethyl sulfoxide undergo differentiation, produce hemoglobin, and lose tumorigenicity. Friend postulates that the mechanism of malignancy is a defect in the ability of cells to differentiate. Thus, like teratocarcinomas and other solid tumors, the leukemias are also differentiating systems that form caricatures of the normal tissue.

Ultrastructural manifestations of differentiation in tumors

Most ultrastructural studies of neoplasms have been made with the idea of establishing the histogenesis of undifferentiated tumors. Pierce (1966) found seminoma (a carcinoma of the testis) to consist of an ultrastructurally heterogeneous collection of cells. At one extreme were cells, presumably stem cells, with undifferentiated cytoplasm characterized by a predominance of polysomes and free ribosomes. At the other extreme were differentiated cells with well developed Golgi complexes containing proacrosomal granules. Mitochondria were sometimes arranged in coiled patterns reminiscent of the mid-piece of sperm. There were few polysomes and ribosomes in these cells, which were often connected by intercellular bridges of the type observed between spermatids. Between these two extremes, cells showed intermediate degrees of differentiation. Influenced by the observation that some of the stem cells of teratocarcinomas could differentiate, the spectrum of differentiation observed in seminoma was considered to be that of a developing system that caricaturized spermatogenesis.

Many investigators, influenced by classical dogma of oncology, have observed similar ranges of differentiation in tumors. For example, Linn *et al.* (1969) examined nasopharyngeal carcinomas with the electron microscope and identified four types of cells. Some were well differentiated and comparable to the cells of normal tissue, but others were less differentiated. The lack of differentiation was attributed, by Linn *et al.*, to variable degrees of dedifferentiation during carcinogenesis. However, the cells of the tumor formed a spectrum of differentiation that again could be considered a caricature of the normal tissue.

Ultrastructural studies of adenocarcinoma of the breast of mice (Wylie, Nakane, and Pierce, 1973) provided information on the state of differentiation of a population of nondividing cells reported by Mendelsohn (1962). Mendelsohn perfused tumor-bearing animals with

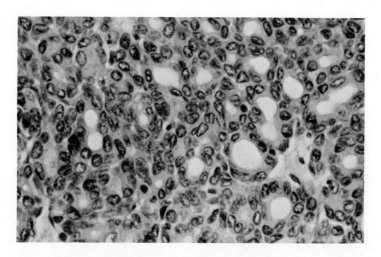

Fig. 3–11 A light micrograph illustrating the features of the best differentiated part of the spontaneous adenocarcinoma of the breast that was given a 2-hour pulse of tritium-labeled thymidine. Note the well formed acini and solid masses of neoplastic cells (450✕). From Wylie, Nakane, and Pierce (1973), courtesy of Springer International.

[3]H-labeled thymidine for a period of 5 days. Approximately one-half of the tumor cells did not synthesize DNA during this time, and he considered them to be stem cells arrested in G_o (see cell cycle, p. 72). Ultrastructurally, as in seminoma, a spectrum of cells that could be arranged in increasing orders of complexity of cellular differentiation was observed (Wylie, Nakane, and Pierce, 1973). At the one extreme were cells containing myriads of polysomes and ribosomes, with few elements of rough endoplasmic reticulum or Golgi complexes. At the other extreme were cells containing secretion droplets, complex arrays of rough endoplasmic reticulum, well developed Golgi complexes, and many kinds of membrane specializations. Intermediate stages of differentiation were observed between the extremes (Figs. 3–11 and 3–12).

Three tumor-bearing animals were perfused with ^{3}H-labeled thymidine for 5 days. Autoradiography with the electron microscope was performed, paying particular attention to the degree of differentiation of unlabeled cells. Other animals received a pulse of ^{3}H-labeled thymidine, and after 2 hours, autoradiography with the electron microscope was performed to relate the state of differentiation to DNA synthesis. It turned out that 11%, 14%, and 35% of the unlabeled cells of the three tumors, respectively, had attained a state of differentiation incompatible with synthesis of DNA. Thus, it was concluded that these cells were postmitotic as a result of differentiation. The balance of unlabeled

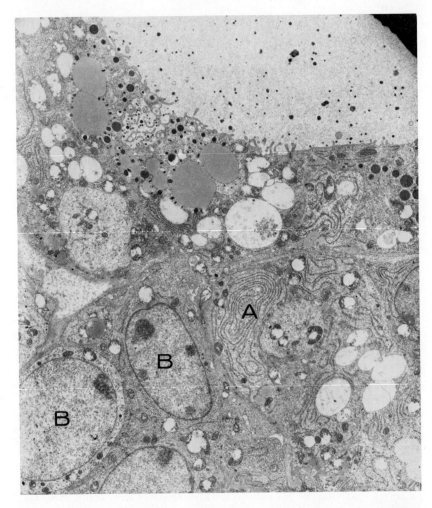

Fig. 3–12 Electron micrograph of a spontaneous adenocarcinoma of the breast showing portions of acini to illustrate cellular details. The cells lining the acinus have microvilli, terminal bars, secretion granules of two types, well developed Golgi complexes, and rough endoplasmic reticulum. Contrast this appearance to the cell labeled A that has many membranous organelles but no secretion granules. Cells labeled B are extremely undifferentiated, and presumably they are stem cells capable of differentiation (7,000✕). From Wylie, Nakane, and Pierce (1973), courtesy of Springer International.

cells had degrees of cytoplasmic differentiation compatible with synthesis of DNA and were presumed, in confirmation of Mendelsohn's notion, to be stem cells in G_0. If not in G_0, these cells were cycling very slowly.

Malignant (Mao, Nalcao, and Angrist, 1966), premalignant, and hyperplastic prostatic tissue have been examined with the electron microscope and cells of Grade 1 adenocarcinoma were found to be indistinguishable from those of normal or hyperplastic tissue (Kircheim and Bacon, 1969; Kircheim, Brandes, and Bacon, 1974). As in the case of adenocarcinoma of the breast, Fisher and Jeffrey (1965) found some of the cells of prostatic carcinoma to be differentiated to the same degree as normal cells. Kastendieck, Altenähr, and Burchardt (1973) observed three types of cells in prostatic adenocarcinomas that they described as proliferative, actively secretory, or transitional between these two. Thus, it would appear that there are undifferentiated, partly differentiated, and differentiated functional cells in the same tumor.

The situation is less clear with mesenchymal tumors, although Friedman and Bird (1969) describe stages in differentiation from embryonal myoblasts to long striated muscle fibers in a rhabdomyosarcoma. We studied a transplantable chondrosarcoma of the golden hamster ultrastructurally. Some of the chondrocytes of this tumor were found to be surrounded by massive amounts of cartilagenous matrix while others were loosely packed and had little matrix. When these cells were examined with the electron microscope, however, a surprisingly narrow range in differentiation was observed among them. All appeared well differentiated. Two hours after a pulse of ^{3}H-labeled thymidine, labeled cells predominated in highly cellular parts of the tumor with few mitotic figures in the chondrocytes surrounded by large amounts of matrix. Ninety-six hours after administration of the pulse, labeled cells were also found embedded in large amounts of chrondromucoprotein. These observations suggest that proliferating elements of the tumor, although very well differentiated, may be capable of synthesizing little in the way of specialized products during their proliferative phase. Some of their progeny, however, become highly functional, presumably an indication of further differentiation. It is not known if chondrocytes become senescent.

The situation in hepatomas * may be similar to that of chondrosarcoma. From an ultrastructural standpoint there is a remarkable degree of morphological homogeneity in minimal deviation hepatomas (Hruban and Rechcigl, 1965) and in human tumors (Svoboda and Higgenson, 1969; Toker and Trevino, 1966). Thus, one can raise the question of whether or not there are stem cells in these minimal deviation hepatomas or whether they behave as the cells of chondrosarcoma.

---

* See p. 183 for a clinical discussion of hepatoma.

Concept of neoplasms

It is now possible to provide a concept of neoplasms more meaningful than the traditional "mass of tissue composed of cells of the body that cause disease." Each of the tumors described in this chapter contains stem cells that proliferate and some of their progeny differentiate. In the case of teratocarcinoma, differentiation evolves normal cells with the capacity to participate in the development of apparently normal tissues and animals. In neuroblastoma, squamous cell carcinoma, and leukemia, the process of differentiation evolves benign cells that are postmitotic and functional.

What is apparent in these four tumors is a natural history in which malignant stem cells eventuate senescent progeny through apparently normal developmental means. The end point for some of the progeny of malignant stem cells is exactly the same as the end point of some of the progeny of normal stem cells. Thus, the teratocarcinoma can be regarded as a caricature of the process of embryogenesis (it even makes embryoid bodies), and the squamous cell carcinoma, neuroblastoma, and leukemia as caricatures of tissue genesis. Evidence obtained from electron microscopy of many tumors is compatible with this interpretation.

The idea that malignant stem cells may evolve normal progeny raises an interesting question: "When in the process of differentiation is the malignant phenotype suppressed?" In the squamous cell carcinoma, it appeared to be suppressed at the time the cells migrated into pearls, because pearls were not tumorigenic. In the breast, cells with the organelles for synthesis of milk could still synthesize DNA and were presumably still malignant, but those containing milk secretions could not and were senescent (Wylie, Nakane, and Pierce, 1973). In the teratocarcinoma, the malignant phenotype was probably suppressed at the time of the first morphological evidence of differentiation into cells representing the three germ layers. In all of the transplants of this tumor made whether from embryoid bodies, small bits of teratoma, or single cells, we have never observed a pure brain tumor, muscle tumor, or pure adenocarcinoma. The conclusion may be drawn that the malignant traits of a tumor are restricted to *stem cells and possibly to some partially differentiated progeny.*

FOUR

Structure
and Biochemistry
of Neoplasms

Introduction

Electron microscopy and biochemistry have had in common the idea that the cause of cancer would be either visible as an ultrastructural lesion or be detectable as a biochemical alteration in some cellular component. In each discipline there are subspecialists with expertise in the investigation of a particular organelle or biochemical pathway. These subspecialists have examined neoplasms from a particular point of view, making morphologic or biochemical comparisons with normal tissue. Some have used adult tissues as controls; others have used embryonic, fetal, or regenerating tissues. As described in Chapter Three, both non-neoplastic and neoplastic tissues are heterogeneous. The critical portion of a growing tissue is the stem cell component, and the fraction of a tissue that is made up of stem cells varies greatly from tissue to tissue and from tumor to tumor. Any comparison of neoplastic and non-neoplastic tissues must take such heterogeneity into account and, ultimately, comparisons, if they are to be meaningful, should be made between neoplastic and non-neoplastic stem cells. Few such studies exist and, instead of reviewing the vast literature of histology and the electron microscopic and biochemical data pertaining to tumors, we will discuss selected aspects, particularly those that pertain to the concept of tumors outlined previously.

Histology of tumors

We are less interested in describing the appearance of a variety
of tumors, or in reiterating the microscopic morphology that distinguishes benign and malignant tumors, than we are in using the lessons
learned from prototype tumors of Chapter Three to understand better
the light microscopic appearance of neoplasms.

The proportion of stem cells to well differentiated elements varies
among tumors and the parts of a tumor. In a given tumor, stem cells
may dominate, resulting in such an undifferentiated appearance that
it is difficult to identify the tissue of origin. In others, the well differentiated elements may so predominate that the tumor closely resembles
normal tissue. As long as some malignant stem cells are present, however, the tumor will behave in malignant fashion. The proportion of
poorly differentiated malignant stem cells to differentiated derivatives
correlates with prognosis in terms of rapidity of growth, invasion, and
metastasis.

In addition to the cellular heterogeneity imposed upon tumors as
a result of differentiation within a stem line, there are multiple stem
cell populations in tumors, each with its own capacity for differentiation. Makino (1956) demonstrated as many as seven stem cell lines in
a murine leukemia, each identified by chromosomal markers. When
teratocarcinomas of mice were cloned, at least six different stem cell
lines were identified. Each was distinctive with regard to growth rate,
number and kinds of differentiated tissues present, and in their ability
to produce embryoid bodies when converted to the ascites (Kleinsmith
and Pierce, 1964). Cloning of a tumor excludes the heterogeneity imposed by the presence of many stem cell lines, but it does not reduce
the heterogeneity that results from the capacity of each clonal line to
differentiate. This point is further illustrated by studies of a malignant
melanoma * that proved that this tumor consisted of several stem cell
lines with remarkably different growth rates and degrees of pigmentation. At one end of the spectrum was a clone that gave rise to slowly
growing, densely pigmented tumors; at the other end of the spectrum
were rapidly growing, amelanotic tumors; and in between were tumors
that were intermediate in growth rates and degrees of pigmentation.
The parent tumor occasionally grew an amelanotic focus or metastasized as an amelanotic tumor. Once amelanotic, the tumor never re-
verted and it synthesized melanin. Although the tumor was carried
for many generations, the slowly growing, densely pigmented com-

---

* See Appendix, p. 197 for a clinical discussion of melanoma.

ponent was not lost. The phenotype of the melanoma was the aggregate of a heterogeneous collection of phenotypes.

It is interesting to speculate on the factors that allow mixed populations to grow with so little segregation. Since all biological phenomena are screened by the law of survival of the fittest, one would expect the phenotype of a tumor to change continually as those stem cell lines best able to survive outgrow the others. One would expect the rapidly growing amelanotic cells to overgrow quickly the slowly growing pigmented ones and evolve a homogeneous amelanotic tumor. Instead, under normal circumstances the tumor remains heterogeneous. When vigorously stressed, as in ascites conversion, the tumor segregates. Appreciation of the cellular heterogeneity of tumors raises more questions than it solves. It stresses the difficulties in maintaining transplantable tumors as models of the original tumor, and it emphasizes difficulties in identifying suitable controls for experiments. These have important implications for the concepts of progression and autonomy (p. 20). Most importantly, the social relationshps of the cells and the reasons why cells with such disparate qualities can remain unsegregated must be determined.

Although organogenesis is present in certain neoplasms, such as teratocarcinomas and Wilms' tumors (Appendix, p. 191.), it is not seen in most malignant tumors. This is to be expected because organogenesis requires the complex interaction of multiple tissues, and since carcinogenesis usually involves only one of the cell types of an organ, there would be no reason to expect organization. For example, an adenocarcinoma of the stomach arises in the mucosa, incites a stroma, proliferates, and forms a mass of neoplastic mucosal cells. The tumor cells may have some capacity to form glands (tissue genesis), but interaction with muscle and other gastric tissues to form a stomach-like organ never occurs. Organization is best illustrated in embryoid body formation in teratomas. Wilms' tumor of the kidney forms imperfect glomeruli, tubules, and various mesenchymal tissues indigenous to the kidney. Certain tumors of the skin have the capacity to form squamous pearls. At the margin of the pearl there is a basement membrane that would normally separate the basal layer of the epithelium from the dermis. At this point, organization breaks down in the squamous cell carcinoma, and the differentiated elements are never supported by the equivalent of a dermis with a vascular supply, lymphatic drainage, or nerves. Thus, there is tissue genesis but no organogenesis.

There are so many examples of the lack of organization and the lack of tissue genesis in tumors that it is useless to recount them. In the case of carcinoma *in situ* in the skin, small aggregates of neoplastic cells, which appear highly malignant, are confined to the epithelium and proliferate and differentiate as evidenced by numerous mitotic figures and the presence of well differentiated or dyskeratotic cells scattered randomly through the neoplastic tissue. Thus, cytodifferentiation occurs in this tumor, but not in a synchronized manner leading to tissue genesis. Whether this effect is dependent upon the lack of communication or transfer of information among neoplastic cells or upon an inability to utilize this information has not been determined. Wessels (1963) has postulated that soluble factors emanating from the dermis play a role in the control of normal epidermal renewal. Chalones may also play a role (p. 142). It is conceivable that a preponderance of undifferentiated cells may create an environment not conducive to tissue genesis, much less organization.

Normal epithelial organs that constantly renew themselves are generally organized in ways that allow removal of senescent cells from the organism. Senescent cells from the gut are sloughed into the fecal stream, cells from the skin are removed as dandruff, and cells from the lung are lost by coughing. In the breast, senescent cells are sloughed into the acini, transported via the ducts to the nipple, and discharged. Senescent cells are produced in certain adenocarcinomas of the breast, but the normal mode of removal of such cells is lost. They undergo necrosis, and when their antigens are presented to the reticuloendothelial system, they elicit a cellular immune response with accumulation of immunocytes. Cellular immune reactions are also present in seminoma and some adenocarcinomas of the stomach. These tumors have a better prognosis than tumors that lack the immune response. Is the prognosis good because of the immune response (as immunologists might believe) or is it good because the tumor has a good capacity for differentiation and evolves senescent cells?

Electron microscopy of tumors

It was hoped that the high resolution and magnification of electron microscopy would allow the visualization of features characteristic and perhaps unique and essential for neoplastic cells. It was hoped that the cause of cancer would be visible. Unhappily, this was not the case (Oberling and Bernhard, 1961; Bernhard, 1963, 1969; Luse, 1961). In the review by Bernhard (1969) it is noted that "the fine structure of tumor cells is as variable as that of their normal homologues and no specific and universal pattern of cancer has ever been found in 15 years of electron microscopical research." What has been found confirms light microscopic studies, *i.e.*, cancers tend to resemble their tissues of origin. This is not surprising because well differentiated neoplasms contain the molecules characteristic of the differentiation, and these are often found in organelles or in configurations recognizable with the electron microscope. This has formed the basis for using the electron microscope as a diagnostic aid. Since there is no problem in diagnosing well differentiated tumors with the light microscope, the goal is to recognize diagnostic features in poorly differentiated tumors. Thus, there are pathologists searching for means of diagnosing tumors undifferentiated by light microscopy (Gyorkey *et al.*, 1975), and there are cell biologists studying and making discoveries in normal tissues and then making comparisons with the corresponding malignancy. The ultrastructural differences between malignant and normal tissues have all been quantitative rather than qualitative, which might be interpreted as further support of the idea that in carcinogenesis one is dealing with changes in control.

Cellular ultrastructure

Exhaustive descriptions have been made of virtually every tumor and organelle. The significance of most of these observations is not known, and probably many of them are trivial. The nucleus of highly malignant cells is usually irregularly shaped, and it is often deeply indented with a condensation of chromatin along the nuclear membrane. It turns out that these changes are also characteristic of the nuclei of many embryonic cells. Other nuclear components, *i.e.*, perichromatin, fibrils, and granules have no distinctive alterations (Bernhard, 1969). Occasionally glycogen or viral inclusions are found in nuclei of malignant cells and the nucleoli are usually numerous, large, and may show degrees of change characteristic of rapidly growing cells.

Chromosomes, which are frequently abnormal in number and configuration in tumors, have no other distinctive ultrastructural features. In their ultrastructural examination of the mitotic process in rat hepatoma cells, Chang and Gibley (1968) found no alterations in the replication and migration of centrioles, the appearance and disappearance of spindle tubules and kinetochores, or the replication and movement of chromosomes.

Correlated ultrastructural and biochemical studies of normal cells has led to the dogma that rapidly proliferating cells usually have cytoplasmic configurations dominated by numerous polysomes unattached to membranes. Thus, it is believed that these polysomes are responsible for the synthesis of structural materials necessary for cellular reproduction. Highly differentiated functional cells have a paucity of free polysomes; instead, their polysomes are attached to membranes and form the endoplasmic reticulum. It is believed that amino acids are incorporated into "peptides for export" on these membrane-bound polysomes, deposited in the cavities of the rough endoplasmic reticulum, and passed to the Golgi apparatus where the sugars are attached. The completed molecule is packaged into secretion granules in vesicles with smooth membranes and extruded from the cell through communications that form by fusion of the membranes of the condensing vacuole and the plasma membrane (Palade and Siekevitz, 1956; Jamieson and Palade, 1968).

The degradative apparatus of the cell is found in the lysosomes, a system of smooth-walled vacuoles that contain enzymes that degrade materials taken up by the cell in pinocytotic vacuoles. These vacuoles fuse with the smooth membranes of the lysosomes, thereby exposing their contents to the degradative enzymes. Lysosomes are also responsible for degrading unnecessary or obsolete cytoplasmic components (autophagocytic vacuoles) (Novikoff, 1961; Novikoff, Essner, and Quintana, 1964).

The energy requirements of the cell are supplied by the enzymatic reactions of the mitochondria. The fine structural relationships of these enzymes to the membranes of the mitochondria and the configurations of the mitochondria in various kinds of cells have been described in detail (Parsons, 1963). Similarly, microfilaments and microtubules are present in the cytoplasm of cells and may be responsible for cell locomotion and cell shape (Porter, 1954; Murphy and Tilney, 1974; Rubin and Weiss, 1975).

Since histologists have long known that tumors appear *undifferentiated* in relation to normal tissue, it should come as no surprise that these attributes are reflected in the ultrastructure of neoplastic cells. Thus, in rapidly growing tumors, the majority of cells have a profusion of polysomes unattached to membranes, at the expense of rough endoplasmic reticulum and well developed Golgi complexes. Lysosomes may be increased in number, but they are often inconspicuous. Autophagosomes are often more numerous than in normal tissue, and phagocytosis of cells is not uncommon in tumors. Necrosis of small numbers of cells is frequently seen. The cells undergoing degeneration are usually well differentiated, and one wonders whether or not some of them are undergoing degeneration because of senescence. Mitochondria are sometimes decreased, normal, or increased in number, and their configuration may vary. Various types of viral inclusion may be present and microfilaments and microtubules vary in number. For that matter, in any given tumor, an organelle or component may be increased or decreased in amount, altered in shape or size, but in another type of tumor the reverse may be observed. Thus, the importance of the variations described is moot and none can be viewed as either causative or as a *sine qua non* of cancer.

Increasing attention has been paid to the role of cell surface phenomena and intracellular relationships in both normal and neoplastic development. Scanning electron micrographs of tumor cells in tissue culture indicates an increase in ruffles and knob-like protrusions. These alterations in appearance are attributed to alteration of microtubules (Porter, Todaro, and Fonte, 1973). The interface between the cell and its environment can serve as a major level of control of growth, differentiation, and cell movement, and alterations in that interface could affect any or all of these functions (see Transformation, p. 98). Unfortunately, the artifacts of preparation of cells for scanning electron microscopy have not been standardized and interpretation is extremely difficult. Much of the data relating to the cancer cell surface is biochemical or deals with such biological properties as antigenicity and agglutinability. The cancer cell membrane has been described as normal although quantitative changes have been discerned in various membrane differentiations. Tight junctions, gap junctions, and desmosomes vary in number. Loewenstein (1966, 1973) has described connections between normal cells across which electron opaque dyes flow readily. Apparently, these junctions are similar to the low resistant nexuses first described by Dewey and Barr (1962) and elucidated by Revel and Karnovsky (1967). The importance of this type of cellular communication in development is not known. Loewenstein (1973), Sheridan (1970), McNutt and Weinstein (1969), and McNutt *et al.* (1971) have reported a diminution in number of junctions between various kinds of tumor cells. Since Revel, Lee, and Hudsphet (1971) reported that only from 1% to 2% of a fat cell's surface was made up of junctions, suggesting that extensive sampling is required to determine if junctions are present in normal numbers. Moreover, it will be necessary to compare the incidence of junctions between normal and malignant stem cells, since the differentiated cells are no longer a component of the malignant process.

Attempts have been made to see if ultrastructural changes might be present in malignant cells that would correlate with invasiveness. A comparison between normal cervical epithelium and cervical carcinoma indicated the presence of more breaks in basement membranes in cancers and fewer desmosomes. Since these features were also noted in carcinoma *in situ*, they cannot be considered to be characteristic of invasion. McNutt *et al.* (1971) reported fewer junctions at the margins of invading squamous cell carcinomas than in the interior of the tumors.

Most of the changes seen in cancer cells are also present in a variety of other non-neoplastic situations, most notably in accelerated growth, such as in tissue culture cells during the logarithmic phase of growth (McNutt *et al.*, 1971) and hepatic regeneration (Jordan, 1964). Franks and Wilson (1970) and Wilson and Franks (1972) established 24 rapidly growing cell lines derived from normal tissues of mice. Eleven of these lines were *tumorigenic*, *i.e.*, they gave rise to tumors when inoculated subcutaneously into immunologically compatible mice; the other 13 lines were not. The tumors produced were undifferentiated sarcomas. The tumorigenic and nontumorigenic lines were examined with the electron microscope and no significant ultrastructural differences were found between them. The tissue culture lines from which the tumors were derived contained two cell types believed to be endothelial cells and pericytes. It was hoped that the tumors developed from them might have histiotypic markers, but this did not prove to be so. Cornell (1969) made similar observations on cell strains and tumors produced by those cells.

Other studies of chemically and virally transformed cell lines showed no consistent ultrastructural abnormalities. Svoboda and Higginson (1968) studied the effects on ultrastructure of hepatocarcinogenic agents in rats including changes that might be correlated with a specific agent or with the onset of malignancy. Although a wide variety of changes was observed, the authors concluded that "it has been stated that there is no specific electron microscopic change whereby the malignant cell can be distinguished from the nonmalignant. The present experiments indicate that the same is true even prior to the appearance of frank neoplasia." This statement probably is true for neoplasms in general.

Since only quantitative differences of questionable significance have been found between normal and outright neoplastic tissues, it is small wonder that significant changes have not been described in *premalignant* lesions, but the search goes on.

Electron microscopy of tumor viruses

Ultrastructural studies of negatively stained preparations have been helpful in describing the morphologic features of oncogenic viruses. Electron microscopy of thin sections has also been used to detect viruses in tumor cells. In some tumors, viruses are easily visualized, but in others, known to be of viral etiology, viral particles cannot be demonstrated. The presence of virions within a tumor does not prove the tumorigenicity of the particles, since cell-free extracts of tumors containing viral particles are often not tumorigenic in any species tested. Extensive search has been made in human tumors in the hope that the presence of viral particles would help establish a viral etiology. Even when found, the inability to test Koch's * postulate in human beings has rendered the approach relatively nonproductive (Dmochowski, 1966; Dmochowski *et al.*, 1967).

DNA tumor viruses include the papovaviruses (papilloma, polyoma, and SV_{40}), adenoviruses, and herpesviruses. In some situations, particles are visible only if there is permissive replication and they have not been observed in transforming infections. In the wart caused by the Shope papilloma virus, which is similar to the human wart, the virion is not detected in proliferating basal cells but is found in crystalline arrays in the nuclei of keratinizing cells prior to cell death.

RNA tumor viruses have been observed in both tumors and normal tissues and the corresponding viral particles are visible after negative staining *in vitro*. The RNA viruses are classified as type A, B, or C, depending upon their morphology, type of maturation, and budding from the cell.

The C type particles have been of particular interest. They develop as crescents beneath membranes and then bud from the cell surface or into intracellular vacuoles as spherical structures with a central dense inner nucleoid. Particles of this type have been found in both animal and human tumors and can be induced in certain cells by application of various substances including chemical carcinogens, x-rays, dimethylsulfoxide, or even by amino acid starvation. Stewart, London, and Lovelace (1964), Dmochowski *et al.* (1967), and Morton, Hall, and Malmgren (1969) have observed the presence of C type particles in short-term cultures of human tumors. Stewart *et al.* (1972) have induced the appearance of these particles in a cell line from a bronchial carcinoma by chemical treatment, but it is not known whether the viruses are passengers or are causally related to the tumor. Of interest is the observation that C type particles in the cells of one species may never be casually associated with production of tumors in that species, but they may often be oncogenic when transplanted into another species.

B type particles were observed by Dmochowski in 1954 in the cytoplasm of mouse mammary tumors, and in 1971 Moore *et al.* found viral particles in human milk that were morphologically similar to the mouse mammary tumor virus. The relationship between the finding of particles in human milk and breast cancer has not been determined.

---

* Koch's postulate: To prove that an infectious agent causes a lesion, the agent must be isolated from the lesion, produce a similar lesion when injected into another animal, and the agent must be extracted from that lesion.

Biochemistry of tumors

The tumor cell has been extensively studied for biochemical uniqueness. Most notable in these studies is a marked similarity in most of the biochemical characteristics of neoplastic and normal cells. The differences that have been described cannot fully explain either the behavior of tumor cells or the events in tumorigenesis. The biochemical alterations do not correlate well with the initiating agent. Some of the changes seem to develop long after initiation and may represent *progression* of events. Nevertheless, it is hoped that biochemical understanding of tumor cells will lead to specific chemotherapy.

Carbohydrate metabolism has been intensively investigated since Warburg (1930) and his colleagues observed that neoplastic tissue slices, incubated *in vitro*, increased lactate formation, even under aerobic conditions. For many years this aberration was considered to be the essential biochemical lesion in neoplasms. More recently, it has been determined that the amount of aerobic glycolysis in both malignant and nonmalignant tissues correlates with the growth rate instead of neoplasia. Apparently, during logarithmic growth all tissues use aerobic glycolytic pathways to produce energy. Neoplastic tissues *in vivo* also have a high rate of lactate formation, but by increasing the pO_2, lactate production can be decreased and the activity of the tricarboxylic cycle increased. No significant impairment of the tricarboxylic cycle has been discovered in tumors (Shapot, 1972; Wenner, 1975).

Slowly growing hepatomas have low glycolytic activity whereas rapidly growing ones have increased glycolysis. Further studies suggest that low levels of pyruvate kinase (PK) in slowly growing tumors favor the use of adenosine diphosphate (ADP) in oxidative metabolism and the high levels of PK found in rapidly growing tumors favor the use of ADP in glycolysis. The high glycolytic activity of some tumors may be due to low levels of ADP acceptor systems in combination with high PK activity (Schapira, 1973).

As for the significance of the Warburg effect, possibly tumor cells or rapidly growing cells do not always have the best contact with a vascular supply and O_2 and adapt to these adverse conditions by using increased glycolysis. Large amounts of glucose are required to supply the energy demands of tumors because of the relative inefficiency of glycolysis in relationship to oxidative metabolism.

Studies of carbohydrate metabolism have stimulated extensive research on enzymes involved in glucose metabolism. In several instances the molecular form of these enzymes differs from the isozymic form found in the corresponding normal tissues. Isozymic differences may be due to epigenetic modifications of the protein molecule (Markert, 1968). These isozymes can be separated by electrophoresis, and the isozymic patterns obtained from tumors often bear more resemblance to those of fetal tissue than to those of the adult tissue from which the tumor was derived (Fishman and Singer, 1975).

When liver cells divide after partial hepatectomy or form hyperplastic nodules during the feeding of a chemical carcinogen, they contain isozymes similar to those of fetal liver cells. For instance, in adult rat liver, hepatocytes contain type I PK (pyruvate kinase). In regenerating rapidly growing livers and in the livers of animals fed carcinogens, there is an increase in type III PK and a decrease in type I PK. In the hyperplastic and regenerating livers, the enzymes eventually revert to the forms and amounts of the adult.

Isozymes have been studied in hepatomas (Walker and Potter, 1972). Minimal deviation hepatoma cells frequently contain the isozyme pattern characteristic of fetal liver cells. The more deviated hepatomas contain the fetal form of aldolase while the minimally deviated hepatomas contain aldolase B. Potter (1964) has proposed the concept that during carcinogenesis either an undifferentiated stem cell or dedifferentiated cell population proliferates, and these actively dividing cells fail to differentiate fully. Some investigators feel that there are marked biochemical similarities among liver tumors and differences between fetal liver, regenerating liver, and neoplastic liver and that these changes represent a *reprogramming* of gene expression. Such apparent reprogramming might also result from a selective process that occurs when tumors are extensively transplanted or carried in culture (Weber, 1973).

An interesting aberration in enzyme function found in a wide variety of tumors is a qualitative and quantitative change in tRNA methylases (Kerr and Borek, 1973). There is a higher level of methylase activity in fetal liver than in regenerating liver, suggesting that growth rate is not the sole determining factor in methylase activity. The assay is usually performed on crude extracts from cells and measures total activity. It is conceivable that specific methylases are elevated or decreased in tumor cells and that the resultant modification of the tRNAs may produce altered nucleotide structures and sequences. Several tumors have been shown to contain hypermethylated tRNA. Some contain tRNA species that are chromatographically distinct from the tRNAs isolated from control tissues. At present, it is not known whether these tRNAs represent new primary structures or are secondary modifications of tRNAs found in control tissues. No functional alteration in coding response or aminoacylation capacity has been demonstrated for these tRNAs. tRNAs have fundamental roles in protein synthesis and other metabolic pathways, and the specific effect that such alterations may play in differentiation or carcinogenesis is not known (Weinstein, 1969; Borek and Kerr, 1972).

Several studies have attempted to show a difference in the DNA of malignant cells as compared to normal cells, but no consistent differences have been detected. The amount of DNA can vary with tumor cells containing a diploid, hyperdiploid, or hypodiploid number of chromosomes (Shearer, 1971). In general, most tumor cells are aneuploid. The only well-substantiated cases in which atypical DNA is present in tumor cells involve inclusion of viral genomes into the cellular genome (Sambrook *et al.*, 1968).

The transcription of DNA has been analyzed by molecular hybridization of the RNA of tumor cells to DNA. In these experiments, polymers of nucleic acids form hydrogen bonded structures with nucleic acids containing a complementary sequence. For example, the sequence ATATGC would specifically anneal to the sequence

TATACG to form a double-stranded molecule, $\frac{\text{ATATGC}}{\text{TATACG}}$. The double

nucleic acids can be isolated and quantitated by either nuclease treatment or hydroxylapatite chromatography.

No study has demonstrated conclusively that there are altered RNA species in the transformed cells except in the cases of viral transformation, where viral sequences are transcribed (Benjamin, 1966). Several investigators using competitive hybridization, annealing of the RNA from neoplastic and control cells to DNA, suggest that there are differences in the RNAs of tumor and normal cells (Church *et al.*, 1969). Busch *et al.* (1974) have found some differences in the sequence of ribosomal RNA from Novikoff hepatomas in comparison with normal liver ribosomal RNAs. These alterations in sequence might be due to the fact that the Novikoff hepatoma is a rapidly growing tissue while the control normal liver is not.

A variety of enzymes involved in nucleic acid biosynthesis such as TK (thymidine kinase) and ribonucleotide reductase have altered activities in transformed cells (Bresnick, 1974). The significance of these changes is not understood. DNA polymerase has also been studied in transformed cells. It has been suggested that this enzyme is more variable in transformed cells and may, by producing errors during DNA replication, lead to changes in DNA composition (Loeb, Springgate, and Battula, 1974). The role of enzymes in the repair of DNA and in the causation of neoplasia will be discussed in the section on genetics.

Because the structure of chromatin is thought to regulate the expression of the information encoded in the DNA, histones and nonhistone proteins from transformed tumor cells and nontransformed tissues have been compared. The histone composition of several solid tumor lines appears similar but differences in histone (H1) composition between rat liver and a hepatoma have been observed (Sporn and Dingman, 1966). The nonhistone proteins are exceedingly difficult to study because there are so many of them. Changes in the patterns of the nonhistone proteins have been observed in certain hepatomas and in dimethylnitrosamine transformed baby hamster kidney cells (Yoeman *et al.*, 1975; Wilson *et al.*, 1975; Fosger, Choie, and Friedberg, 1976).

The production of proteolytic enzymes by tumor cells may determine some of the phenotypic characteristics of malignant cells (Roblin, Chou, and Black, 1975). In 1925 Fisher noted that cultures of tumor cells could lyse fibrin clots and that this effect was dependent upon the type of serum used in the culture medium. Reich and his coworkers (1974) described increased fibrinolytic activity in cell cultures of virally and chemically transformed cells. They placed either the lysed cells or the fluid harvested from cells on plates containing radioactive fibrin and measured the counts released in the media. Fibrinolysis was due to a plasminogen activator that cleaved plasminogen to plasmin. Plasminogen activator activity also increases in chick embryo fibroblasts infected by Rous sarcoma virus (RSV). This increase was detected within hours after a cell infected with a temperature sensitive mutant Rous sarcoma virus was shifted from a temperature nonpermissive for transformation to one permissive for transformation (see p. 108). This effect was reversible. These studies also suggest that the expression of some of the properties of transformed cells, such as a rounding of the cells, increased cell migration after wounding of a culture, and growth in agar might be related to the presence of plasminogen activator and plasminogen (free of inhibitors) in the culture medium. The ability of transformed cells to grow as anchorage independent cells (*i.e.*, the ability to grow in semisolid support) and their tumorigenicity have been correlated with their plasminogen activator activity (Pollack *et al.*, 1974). Mouse melanoma cells lose their tumorigenic potential when treated with bromodeoxyuridine. Concomitantly, their plasminogen activator activity decreases (Christman *et al.*, 1975).

There are many tissues that normally contain high levels of plasminogen activator or proteolytic enzymes and in some situations the untransformed cells appear to have higher levels of plasminogen activator activity than the transformed cells (Mott *et al.*, 1974). There are *in vivo* experiments that suggest that the level of plasminogen activator in cells may be modulated by factors in the host animal (Jones, Lang, and Benedict, 1975).

Human tumor fragments, cells isolated from explants of human tumors, and cell lines from tumors frequently have high levels of plasminogen activator activity (Rifkin *et al.*, 1974). Many breast cancers are fibrinolytically active and there is some correlation between high fibrinolytic activity in mammary carcinomas and invasion and metastasis.

Proteolytic enzymes have been implicated as growth stimulants (Burger, 1970), and it has been suggested that they may be involved in tissue vascularization, fibrinolysis, and tissue remodeling (*e.g.*, wound healing or involution of the lactating breast). The proteolytic enzymes may be extremely important in invasion of tumor cells through basement membranes that border epithelial surfaces and blood vessels.

Protease inhibitors such as N-α-tosyl-L-arginine methylester, which inhibits trypsin, have been shown to cause a cessation of growth of transformed cells in tissue culture (Schnebli and Burger, 1972). The final density of the treated transformed cells is similar to treated control cells. Proteases may cause modifications of the structure of the cell plasma membrane. For example, trypsin treatment of cells in culture appears to cause a burst of cell division and also changes the agglutinability of cells. It has been speculated that protease production by transformed cells may be related to alterations in the plasma membrane of transformed cells.

In the past several years there have been intensive studies of the chemistry of normal cell surfaces (Hynes, 1976; Nicolson, 1976). The plasma membrane appears to be made up of a central phospholipid bilayer with globular proteins either embedded in the lipid exposed on the surface or extending completely through the membrane with exposure on both internal and external surfaces. Many of these proteins are glycosylated. Glycolipids may extend from the lipid layers and present exposed polysaccharides on the surface of the cell. There may be a structural and functional relationship between the membrane glycoproteins and glycolipids and the underlying microtubular and microfilamentous structures of the cell. The membrane appears to be a mosaic structure with capability for movement of its components, *i.e.*, a fluid mosaic (Singer and Nicholson, 1972).

The chemical composition of tumor cell plasma membranes has been compared to normal cells in tissue culture systems. The differences are not always constant and may vary with clonal isolates and cell density. In some tumor cells, reactions for carbohydrates are more intense at the surface, suggesting an increased accessibility of these components. Some cells transformed by DNA tumor viruses contain less total sialic acid (Culp, Grimes, and Black, 1971). In cells transformed by polyoma virus or SV_{40} there is less N-acetylneuraminic acid and N-acetylgalactosamine than in untransformed cells (Meezan *et al.*, 1969).

There are extensive and variable changes in the glycoproteins of tumor cells. Some experiments indicate that the glycoproteins from tumor cells have decreased sialic acid while other experiments indicate an increase in the amount of sialic acid (Culp, Grimes, and Black, 1971). There are contradictory reports of changes in sialytransferase activity in transformed cells (Bosman, Hagopian, and Eylar, 1968; Warren, Fuhrer, and Buck, 1972). Investigators have labeled the membranes of cells with radioactive sugars and chromatographically and electrophoretically fractionated the glycoproteins (Buck *et al.*, 1970). The differences found between transformed and normal cells are frequently as great as those between clones of cells (Fishman *et al.*, 1972). By using lactoperoxidase to iodinate the surface proteins of plasma membranes, it has been demonstrated that there is frequently a loss of specific protein of approximately 250,000 molecular weight after viral or chemical transformation. This is an example of an early membrane change, the significance of which has yet to be determined (Hynes, 1976).

In general, there are a simplification and a shortening of the carbohydrate chains in glycolipids from tumor cells. Along with the apparent incomplete synthesis of the carbohydrate chains there is a decrease in activity of several glycosyl transferases (Itaya, Hakamori, and Klein, 1976).

As normal cells in culture reach saturation density, there is an elongation of the carbohydrate chains of glycolipids. This has been termed an *extension response* (Hakamori, 1973). There is speculation that glycosyltransferases on one cell surface might recognize and glycosylate either the glycolipids or the glycoproteins on an adjacent cell's surface (Roth, McGuire, and Roseman, 1971). It is not known whether modifications of the polysaccharide components of glycolipids or glycoproteins cause the loss of contact inhibition of tumor cells in tissue culture.

From the preceding discussion it is apparent that generalizations about the biochemical changes of membranes of tumor cells must be made with caution. Most of the membrane studies have been made on cells growing in tissue culture and possible artifacts must be considered, including those resulting from removal of cells from plastic surfaces, the phase of the cell's growth, and the percentage of cells actually transformed.

Aub, Tieslav, and Lankester (1963) found that an extract from wheat germ would agglutinate transformed cells at much lower concentrations than those required to agglutinate normal cells. Several lectins have now been isolated from plants that agglutinate animal cells (Sharon and Lis, 1972). These lectins have been used extensively to probe for differences in the components of plasma membranes. They are thought to bind to specific saccharide(s) on glycoprotein(s) or glycolipids located on the cell surface. They include concanavalin A (con A), wheat germ agglutinin (WGA), and soybean agglutinin (SBA). The lectin-cell interaction may be similar to antibody-cell interactions and in some instances may stimulate division of cells. The surface membrane receptor glycoproteins or glycopeptides for several lectins have been isolated (Rapin and Burger, 1974).

Con A appears to bind to tumor cells and normal cells in roughly the same amount, but it agglutinates tumor cells and is more toxic to tumor cells, especially at low concentrations (Ozanne and Sambrook, 1971). Increased agglutination with lectins has been demonstrated for cells transformed by chemical carcinogens, DNA and RNA tumor viruses, and certain nontransforming viruses. Experiments using temperature-sensitive mutants of oncogenic viruses have indicated that in virally infected cells, the viral genome is important in maintenance of the increased agglutinability (Eckhart, Dulbecco, and Burger, 1971). These studies suggest that the surfaces of tumor cells differ from those of normal cells. Nicholson (1971) speculates that the mobility of agglutinin sites on tumor cells or trypsinized normal cells allows the clustering of receptor sites that favors agglutination.

There are several experiments demonstrating that some fetal cells are more agglutinable by lectins than cells from adult tissues (Moscona, 1971). Perhaps the membrane alterations of tumor cells and nonneoplastic undifferentiated cells are similar.

Both the actin cables and the tubules consisting of tubulin appear to be in a depolymerized state in transformed cells (Weber *et al.*, 1974; Brinkley, Fuller, and Highfield, 1975). This is similar to normal cells during mitosis. These cable networks may be responsible for maintaining certain membrane functions and the shape of cells. The levels of specific cyclic nucleotides seem to be related to the polymerization state of the components of the intracellular microtubules and to cell morphology.

Decreased intracellular levels of cAMP (cyclic adenosine monophosphate) have been observed in tumor cells (Sheppard and Lehman, 1973). Adenyl cyclase, which hydrolyzes cAMP, is located at the plasma membrane, where several hormones act to regulate the cAMP levels (Sutherland, 1972). The cyclic nucleotides in turn seem to act as a messenger in eliciting a cellular response. Some investigators postulate that both the intracellular cAMP and cyclic guanosine monophosphate levels are important for the control of cell division and for the differentiation of cells. Experiments have shown that the morphology of neuroblastoma cells in culture or Chinese hamster ovary cells can be altered by adding BU_2-cAMP (N_6,O_6-dibutyryl adenosine 3',5'-cyclic monophosphoric acid), a derivative of cAMP capable of entering cells (Prasad and Hsie, 1971). The neuroblastoma cells appear more differentiated and the Chinese hamster ovary cells appear more like differentiated fibroblasts (Hsie, Jones, and Puck, 1971). The lectin agglutination of some tumor cells treated with Bu_2-cAMP also decreases as the morphology changes. Perhaps the inability of carcinogen-treated stem cells *in vivo* to modulate their cAMP levels in response to hormones precludes normal differentiation while allowing cell division and tumor production.

Cell surface alterations may be detected as antigenic changes. These have been observed on both virally and chemically transformed cells and on cells from human tumors. A common antigenic alteration can be detected in virally induced tumors or cells transformed by the same strain of virus. The oncornaviruses (RNA tumor viruses) contain gs (group specific) antigens (Sarma, Turner, and Huebner, 1964). The gs antigen in chicken cells appears to be determined by an autosomal dominant gene, but there is evidence that the genes of the RNA tumor virus contain the information for the gs antigen (Payne and Chubb, 1968). Avian sarcoma viruses transform mammalian cells and these cells then synthesize the Rous sarcoma virus gs antigen (Huebner *et al.*, 1964). Because the gs antigen for murine oncornaviruses is found in fetal tissues, it has been suggested that the genome for these viruses is present during embryogenesis and may even have a functional role in differentiation and development of the fetus (Huebner *et al.*, 1970).

DNA tumor virus either contains information for or stimulates the synthesis of new antigens on transformed cells. The SV_{40} virus causes the production of a TSTA (tumor specific transplantation antigen) in tumor cells that are detected by tumor rejection in sensitized hosts (Defendi, Lehman, and Kraemer, 1963). This antigenic change is in the surface membrane and is probably coded for by the virus. A new surface antigen called the S *antigen* is also found on cells transformed by SV_{40}, but this antigen is a normal component of uninfected cell membranes. The S antigen can be exposed on normal cells by treatment with trypsin. Thus, the SV_{40} virus has caused a pre-existing antigen to be exposed (Butel *et al.*, 1972). One virus may produce more than one antigenic alteration in the cell surface membrane and some of these changes may be in addition to viral coded products, exposure, deletion, or new synthesis of cellular products found in the plasma membrane.

The transformation of cells by chemical carcinogens involves complex antigenic changes in the cell surface (Prehn and Main, 1957). Experiments have shown that a single chemical carcinogen can cause many antigenic changes in transformed cells (Baldwin, 1973). This differs from viral transformation in which tumors caused by the same virus show common immunological determinants. Some of the antigenic changes expressed in chemically transformed cells are probably fetal antigens.

In certain animal and human tumors there is expression of carcinoembryonic antigens (CEA). In 1965 Gold and Freedman demonstrated the presence of CEA in extracts of tumors of the digestive tract and in the serum of patients with tumors of the gastrointestinal tract. The test uses antiserum obtained from rabbits or goats immunized with antigens extracted from liver metastases of colon carcinomas. This antiserum is reacted with the serum of patients; from 70% to 95% of patients with cancer of the colon have circulating carcinoembryonic antigens. The test is most likely to be positive in patients with metastases. In some patients with primary cancers of the digestive tract an antibody to CEA may be detected. CEA may disappear if the tumor is resected, but it rises if there is reappearance of the tumor or metastasis. CEA is not specific for tumors of the gastrointestinal tract and is found in the sera of patients with inflammatory disease, particularly in ulcerative colitis in which there is extensive proliferation of the mucosal cells of the large intestine, as well as in patients with neoplasms in other locations (Zamcheck, 1975).

Another interesting antigenic change found in human tumor cells is the exposure of the blood group A antigens on adenocarcinoma cells. It has been speculated that the increased incidence of blood group A in people with gastric carcinoma is due to a lack of antibodies to the A antigen. More studies are revealing that in human tumor cells there is an unmasking of antigenic sites that have become cryptic during differentiation.

Alpha-fetoprotein, a major protein produced by the liver in the early fetus, is not normally present in the adult. It is frequently detected in humans or experimental animals with hepatoma, teratocarcinoma, or embryonal carcinoma. It was hoped that detection of an alpha-fetoprotein in the serum would allow early detection of tumors, but as with CEA, it has been found that when there is increased cellular turnover and regeneration, as in hepatitis, the level of alpha-fetoprotein is also increased (Abelev, 1971).

The study of antigenic changes on the surfaces of tumor cells has become an area of intensive research because an understanding of this phenomenon might lead to (1) an understanding of the basic changes relating to the malignant behavior of these cells, (2) an understanding of the relationship of the membrane to cell division, (3) an understanding of the role of the host's immune response in either rejection or nonrejection of the tumor cells, and (4) the development of agents such as BCG that might enhance the host's capability to recognize and reject tumor cells.

The aberrant production of proteins or polypeptides by human tumors is an interesting and important phenomenon in clinical medicine (Lipsett, 1968; Hall, 1974). There are many instances in which tumors of non-endocrine tissues secrete hormones. The most frequently observed example is the production of ACTH, insulin, gonadotropins, and parathormone by bronchogenic carcinomas. The secretion of excessive amounts of these hormones may be the earliest detectable symptoms of the tumor. There may be many as yet undetected examples of tumors yielding products that are not characteristic of what we assume to be the tissue of their origin (*ectopic protein syntheses*). If an uncommitted stem cell were a target for the neoplastic initiating event, the production of hormones could be an example of abnormal differentiation.

In this survey we have selected certain examples to establish a point of view. No attempt has been made to be all inclusive. The generalization that emerges from these examples is that many of the characteristics that have been observed in tumor cells are dissimilar when compared to their corresponding adult tissues but similar to the corresponding fetal tissues. Not all these changes may be related to the growth rate of the tissue and some may be the result of the selection and proliferation of stem or blast-like cells after exposure to carcinogens. The characteristics observed in tumor cells may then represent failure or aberrations of normal differentiation. Thus, it is apparent that true controls for studies on transformed cells are really normal stem cell populations. Meaningful observations will require techniques for the isolation of stem cells.

We can see that the study of tumor cell biochemistry is continually evolving. Exciting areas of research appear every few years. For instance, in the 1930's it appeared as though the understanding of aerobic glycolysis would clearly lead to an understanding of tumor cell biology. These studies led to further studies on the enzymes and isozymes in tumor cells. In the 1960's the analysis of the nucleic acids and the biosynthesis of nucleic acid macromolecules in tumor cells moved to the forefront and the study of tumor viruses became intense. The late 1960's and 1970's saw the emergence of tumor immunology, which led to research on the structure and function of cell membranes.

In each area of research further investigation with more and better controls has put the observations into perspective so that there are no alterations in the biochemistry of tumor cells that can be described as pathognomonic for cancer.

FIVE

Origin of Neoplastic
Stem Cells

Introduction

There have been many theories of the origin of tumors, but it will serve no useful purpose to review them because most were postulated before much was known about developmental biology and when little was known about tumors other than that they were masses of tissue made up of altered cells capable of invading, metastasizing, and causing death by cachexia. These theories have all centered around the stability and heritability of the phenomenon and the lack of differentiation and organization in malignant tumors. Unfortunately, *lack* has been equated with *loss*. Thus, carcinogenic agents are still believed by some to have as their target fully differentiated cells that *lose* some of the overt manifestations of the differentiation; this process has been called *dedifferentiation*. If the dedifferentiative process were extreme, many of the characteristics of the normal differentiation would be lost and an undifferentiated, rapidly proliferating, widely metastasizing, highly malignant tumor would result. If the losses were less extreme, few departures from the normal phenotype would occur with the evolution of less malignant tumor cells. Variations in the degree of dedifferentiation in cell populations would explain the heterogeneity in differentiation observed in a tumor.

Our concept postulates that a tumor is an aberration of tissue renewal that results in a caricature of the parent tissue. Thus, it would seem logical to search for the origin of cancer in the cells involved in cell and tissue renewal. Since cell and tissue renewal are mediated by cell division and differentiation, the strategy will be to review briefly pertinent aspects of those processes and then focus upon the cell of origin of tumors.

Tissue renewal and its modulation

Cell division, differentiation, and organization into tissue are the processes by which the developing organism achieves mature form. *Biological maturity* in human beings is defined as the age at which maximal height is attained. Maturity does not mean, however, that cell division and cell differentiation cease since these are the processes by which tissues undergo renewal. The controls of proliferation and differentiation are regulated to maintain the status quo. There is no increase in mass and yet there is a flexibility that provides a means by which tissue maintenance can be altered in response to environmental signals (Goss, 1972; Cairnie, Lala, and Osmond, 1976).

The specialized tissues of an adult animal contain a population of undifferentiated but already *determined* precursors called *stem cells*. It is by the proliferation of stem cells and the maturation of some of their progeny that differentiated tissues are renewed or restored. The stem cells of a particular tissue are *determined*; that is, under normal circumstances, their progeny will differentiate into that type of tissue and not into any other. For example, the stem cells of intestinal epithelium that differentiate into intestinal epithelium only are present in the crypts at the base of the villi where they undergo cell division. After division, one of the progeny cells remains in the crypt as an undifferentiated stem cell, but the other cell migrates along the surface of the villus and differentiates into one of the highly specialized cells that performs the function of mature gastrointestinal epithelium. Little is known about the selection of which cell remains as a stem cell and which cell differentiates. These differentiated cells do not undergo further division as they migrate toward the tip of the villus. The life of such a cell is approximately 4 days (Leblond and Stevens, 1948) and when senescent, the cells are sloughed from the tip of the villus into the fecal stream. Normally, cell production equals cell loss —a dynamic equilibrium. In addition to gastrointestinal mucosa, epidermis, respiratory epithelium, testicular epithelium, and hematopoietic tissue undergo renewal rapidly and continuously, and the life of the organism, but not the lives of the individual cells, is dependent upon this process.

Other tissues in the mature organism undergo renewal at a slower, sometimes barely perceptible rate, but they retain the potential for rapid growth when the situation requires it (Baserga and Wiebel, 1969). The liver is an example of such a tissue. Turnover of hepatocytes normally occurs at such a slow rate that at any given time it is difficult to find a mitosis in many sections from a liver. If, however, three-quarters of the liver are resected, the remaining tissue proliferates rapidly, and the original tissue mass is regenerated within a matter of a few days.

The mode of tissue renewal has not been worked out for all organs and tissues. For example, it is not known if liver is renewed from stem cells such as those described above in the intestine. There is evidence suggesting that liver cells may never become postmitotic and that renewal of liver occurs from apparently well differentiated cells with their well developed endoplasmic reticulum and Golgi apparatus. Hepatocytes would presumably stop the complex synthesis of differentiated molecules in response to an appropriate environmental stimulus, synthesize DNA, and divide.

Similarly, it is not known whether the cells of mesenchymal tissues have a life cycle similar to those in the gut or marrow or if they resemble liver cells and remain plastic and capable of cell division whenever the need arises. It has long been accepted that mesenchymal tissues remain plastic and can dedifferentiate when the circumstances demand and redifferentiate into the appropriate mesenchymal tissue. This notion receives its strongest support from studies of limb regeneration in amphibians. After a limb has been amputated, muscle, cartilage, and osteocytes supposedly dedifferentiate at the site of amputation to form a blastema. There is no question that a blastema forms, but the evidence that it develops by dedifferentiation is not compelling. The cells of this undifferentiated blastema then proliferate, differentiate into the appropriate mesenchymal tissue, and organize themselves into a new limb. Limb regeneration does not occur in mammals and damaged muscle or cartilage do not regenerate after injury; instead, repair is by fibrosis and scar formation. Moss and Leblond (1970) have shown that striated myotubes contain undifferentiated cells that in the growing rat are capable of mitosis. Some of their progeny differentiate into myocytes. Thus, there is a mechanism for renewal of myocytes, but experience shows that myotubes are not regenerated after injury. Why don't these cells regenerate muscle? The probable answers are that (1) injury to muscle is usually accompanied by damage to the basement membrane and other components of the myotubular framework and (2) an intact framework is required for regeneration to occur. The precedent for this conclusion exists in the kidney. Specific toxins may kill the differentiated renal tubular epithelium without damaging its supporting framework, and under these circumstances the epithelium regenerates from primitive cells. It may well be that a similar population of undifferentiated cells exists in liver, but it is obscured by an overwhelming number of differentiated cells.

In certain situations cell proliferation and differentiation do result in an increase in mass, a phenomenon best illustrated in pathological and experimental situations. If testosterone is injected into a male rodent, there is a significant increase in size and weight of the ventral prostate. This is due to a marked increase in the number of prostatic cells of normal size, shape, and configuration that form additional functional glands. If the injections of testosterone are stopped, growth ceases and the prostate reverts to normal size. This is an example of hyperplasia, a situation in which there is an increase in the number of cells resulting in an increased mass of tissue that persists as long as the inciting environmental stimulus is present. Cyclic hyperplasia normally occurs in the endometrium during reproductive years. The endometrium undergoes hyperplasia in response to estrogens produced during the first part of the menstrual cycle. About halfway through the menstrual cycle, progesterone is produced and causes a morphological change in the endometrium. The uterine mucosa, now increased in mass and changed in its morphological appearance and function, provides an acceptable nidus for a fertilized ovum. However, the endometrium cannot maintain itself in the absence of the endocrine changes accompanying pregnancy and is shed during menstruation. This restores the mucosa to the resting state, and the cycle repeats itself.

A reversible change in the morphologic appearance of a cell or tissue in response to environmental stimuli is called *metaplasia*. Thus, the uterine mucosa underwent reversible morphological change in response to progesterone, a metaplasia. The ciliated, mucous-secreting bronchial epithelium often undergoes metaplasia to squamous epithelium in response to heavy cigarette smoking and other noxious agents. Certain inflammations of the kidney bathe the kidney pelvis in irritants, resulting in squamous metaplasia of the uroepithelium. In response to large amounts of vitamin A, squamous epithelium of chick rudiments *in vitro* undergoes a glandular metaplasia with the production of mucous-secreting cells. The mucous-secreting cells do not differentiate from mature squamous cells; the latter degenerate and the glandular cells differentiate from the progeny of the normal stem cells of the skin which, in the absence of the vitamin, ordinarily would form squamous cells.

Like hyperplasias, metaplasias persist only as long as the inciting environmental stimulus. They too are the equivalent of modulations of development, and they presumaby represent the range of reactivity of normal stem cells to certain environmental stimuli. It must be noted that metaplasias do not cross embryonic germ layers. For instance, if an epithelium becomes metaplastic, it cannot form muscle or cartilage; it can only form a different and simple type of epithelium.

The stem cell population of a tissue is the reservoir of tissue reactivity, whether in the form of tissue renewal, regeneration, hyperplasia, or metaplasia. It is important to emphasize that one of the fundamental differences between hyperplasia, metaplasia, and neoplasia is that hyperplasia and metaplasia persist only as long as the inciting stimulus is present. The neoplastic change is permanent and is inherited by succeeding generations of cells in the absence of the original stimulus. Thus, certain environmental stimuli modulate cellular controls of proliferation and differentiation, but those resulting in neoplasia go beyond reversible limits. It is the explanation for this heritability and stability of neoplastic change that has resulted in controversy.

Cell cycle

Until the 1950's analyses of cell division and tissue growth were limited to counting mitotic cells and calculating their percentage in a tissue culture plate or tissue section. With the introduction of tritiated thymidine and autoradiography (Doniach and Pelc, 1950) it became possible to characterize the kinetics of cell proliferation more precisely. The cell cycle (time between two succeeding cell divisions) and its component phases have been characterized biochemically and found to include a number of critical events that occur in a specific chronological sequence (Mueller, 1969; Tobey *et al.*, 1974). Some investigators have attempted to distinguish events that might be crucial to neoplastic transformation, while others have sought those events that might allow increased susceptibility to agents that might prevent the neoplastic cell's proliferation or that even might kill it (Baserga, 1971; Warwick, 1971). Some chemotherapeutic agents act at specific points in the cell cycle, and a combination of such agents, given at suitable times, might provide maximal effect. This is the basis for the complex schedules of combined therapies (chemotherapeutic and/or radiation) that currently are being used (Chapter 11).

The methodology, observations, and concepts of cell cycle have been reviewed (Baserga, 1971). As usual, most of the data have been derived under conditions in which variables can be minimized, *i.e.*, *in vitro*. Under optimal growth conditions, many normal or neoplastic or plant cells have a similar interval between successive cell divisions, averaging from 15 to 24 hours. *In vivo*, cell cycle time varies from tissue to tissue, reflecting, among other variables, the heterogeneous nature of a tissue. In the murine gastrointestinal tract, for example, the mean cell cycle times are 181 hours in the esophagus, 17 hours in the ileum, and 36 hours in the colon.

The cell cycle, with minor variations, appears to be the same for all eukaryotic cells and is divided into the following phases (Fig. 5–1). G_1 begins with the termination of the preceding cell division (in rapidly growing populations) and involves preparation for DNA synthesis, S is the phase of DNA synthesis, G_2 is the interval from the termination of DNA synthesis to the onset of division, and M is the phase during which mitosis occurs. Mitosis, the only phase that can be discerned by routine light microscopy, comprises less than 5% of the cycle of most proliferating cells. The times required for S, G_2, and M remain constant regardless of the overall length of the cycle. G_1 is the variable phase, lasting days, weeks, or even months in slowly growing systems; or it can be barely detectable in some very rapidly proliferating systems.

By synchronizing cells, *i.e.*, using a variety of techniques to obtain a population in a particular phase, and then stimulating their proliferation, sufficient homogeneity has been obtained to characterize the sequence of cyclic events. Sequential synthesis of RNA and protein has been found, and the patterns of synthesis of groups of proteins and of particular proteins have been determined (Petersen, Tobey, and Anderson, 1968). In particular, the enzymes involved in DNA synthesis have been identified, and their synthesis during G_1 has been characterized (Mitchison, 1969). Less specific information is known about the

Fig. 5–1 Cell cycle: G_1, termination of the preceding cell division; S, phase of DNA synthesis; G_2, interval from the termination of DNA synthesis to the onset of division; and M, phase during which mitosis occurs.

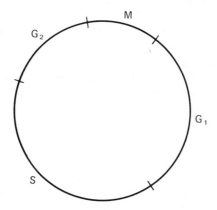

biochemical events that occur in S, G_2, and M. It is known that during G_2, for example, proteins that are necessary for mitosis are synthesized, and by blocking protein synthesis up to 15 minutes before the onset of mitosis, one can interfere with cell division. The specific identities of these proteins and their functions are essentially unknown (Mueller, 1969).

Even less is known about the signals that act to trigger the successive steps of the cell cycle or about the controls that modulate these signals (Padilla, Cameron, and Zimmerman, 1974; Clarkson and Baserga, 1974). A key event occurs after cell division, an event that determines whether or not the new cell will continue to cycle mitotically or will enter a nonproliferative phase. The latter situation, termed G_0, takes the cell temporarily, or permanently, out of the cycle. The cell maintains its appearance as an undifferentiated stem cell (Wylie, Nakane, and Pierce, 1973) and may stop cycling and begin specialized functions (differentiate), such as mucin secretion and insulin synthesis, or when the situation demands it may recommence cycling. Some of the reported variations in the length of G_1 may actually represent variations in the time between the termination of the previous cell division and the decision to prepare for a new round of DNA synthesis. In addition, variations may also occur between the onset of preparation for DNA synthesis and S (definitive G_1).

The onset of DNA synthesis appears to be triggered by one or more substances demonstrated in cell hybridization studies. Rao and Johnson (1970) have fused cells in S with cells in G_1. The G_1 component of the hybrid begins to synthesize DNA. Initiation of DNA synthesis is believed to be the result of a triggering substance that originates in that part of the hybrid cytoplasm contributed by the "S" phase cell. Hybridization of cells in mitosis with G_1 or G_2 cells results in condensation of their chromatin, a hallmark of prophase; again, a substance originating in the M cell component is invoked as the triggering mechanism.

Prescott (1976) has postulated a control system that includes a protein that initiates DNA synthesis. This initiator protein is controlled (at the level of transcription) by a repressor-corepressor complex. The repressor protein remains within the cell, but the corepressor can leave the cell. The pool of corepressor is labile, and it is postulated that the amount that reenters the cell will determine the amount of DNA initiator synthesized; breakdown in such an autoregulatory mechanism is suggested as a mechanism for the onset of cancer.

The rate at which any tissue grows is determined by three factors: (1) the rate of proliferation of cells, (2) the number of cells engaged in cell division, and (3) the loss of cycling cells. The rate of proliferation and the number of cells in the process of dividing represent the theoretical rate of growth. Since tissues do not accrue in mass at this theoretical rate, mechanisms have been sought to explain the discrepancy between the calculated and the actual rates of growth. It has been discovered that some cells may stop cycling as the result of conversion to G_0 stem cells, cell differentiation, or cell necrosis. This is a matter of concern to chemotherapists who frequently use drugs that are more efficient as cytocidal agents at specific times in the cell cycle. As a result, there have been extensive studies on the cell cycle of tumor cells, but none of these has found any abnormal features when compared to those of normal cells.

Certain parts of the cell cycle appear to be more vulnerable than others in neoplastic transformation and in initiation by chemical carcinogens. It has been suggested, in studies using synchronized cells, that the cell is most vulnerable to transformation in either late G_1 or early S. In addition, fixation of transformation, which imposes on the cell the ability to remain transformed and to transmit the phenotype to progeny cells, requires a round of DNA synthesis. This appears to be directly analogous to the situation in differentiation. When cells are treated with carcinogens and are not allowed to divide, the number of transformed cells is markedly reduced. This may be the experimental equivalent of the well-known clinical fact that neuronal cells, which after the age of 2 years are incapable of mitosis, never give rise to tumors in the adult.

Cell differentiation

Differentiated cells are characterized by specialization of function that is not in itself essential for the life of the cell, but it is essential for the well-being of the organism. Thus, the organism is dependent upon the successful development and integration of its specialized component cells and tissues, and it is this specialization that distinguishes mammalian cells from bacterial cells. The fertilized egg has the potential, given the appropriate environment, to produce all of the differentiated cells of the body, including the germ line, which is differentiated for continuation of the species. The fertilized egg is thus totipotent, but its immediate fate is to divide into two cells, blastomeres, which in turn divide again. If these blastomeres are suitably cultured, they, too, are totipotent and capable of forming a new animal given the appropriate environment. Eventually some blastomeres gain special attributes, become different from each other, and are no longer totipotent. Their potential has been restricted by the process of becoming different. The newly acquired specialized phenotype is heritable and stable, and the cell will not revert to the undifferentiated state under normal circumstances. These are the essential characteristics of a differentiation.

As blastomeres accumulate in the rodent, trophoblast differentiates. In other words, the potential of these cells is reduced to the development of placental tissue that will ultimately nourish the embryo. Meanwhile, other blastomeres are destined to become the embryo proper and form the embryonic epithelium. A layer of endodermal cells differentiates from and surrounds this epithelium. The distal part of the endoderm will form the yolk sac. At this stage, the rodent embryo is made up of a small nidus of embryonic epithelium overlain by a layer of proximal endoderm. Mesoderm forms between the two (Bonneville, 1950).

The first mesodermal cells arise by polarization of embryonic epithelium, but whether or not specific inducing molecules synthesized by endoderm are required as some believe is not known. Possibly some of the embryonic epithelial cells express a potential in the presence of a suitable environment nonspecifically created by both endodermal cells and embryonic epithelium. Possibly the reaction is mediated by cell-cell contact.

At this stage the mouse embryo consists of an inner core of embryonic epithelium (develops into brain, skin, glands, etc.) overlain by mesoderm (develops into muscle, cartilage, bone, connective tissue, etc.) and endoderm (develops into trachea, lungs, cells lining the gastrointestinal tract, etc.). Each tissue is added in order, and each presumably contributes molecules to, and thereby changes, the microenvironment required for the expression of yet other phenotypes. These sequential changes increase specialization at the expense of potential so that the net result of development is really the orderly phasing out of potential. This process is divisible into two aspects, determination and expression. In determination, stable populations of stem cells are produced, each restricted to a particular differentiation although not necessarily containing the molecules characteristic of the differentiation. These stem cells are capable of dividing, with one progeny cell remaining as a determined stem cell and the other expressing the differentiated state. Regulation at this level is the mechanism whereby mature, adult tissues are maintained in the status quo but with enough modulating potential to respond to the vicissitudes of the environment.

The specifics of determination are not known, but they may depend upon signals in the environment. For example, some believe that an inductive signal generated by endoderm causes polarization of the underlying embryonic epithelium to form the first mesodermal cells. In similar manner, the optic vesicle, which lies immediately beneath ectoderm of the embryonic head, induces head ectoderm to form a lens. Subsequently, the presumptive lens cells, which now are incapable of forming any other kind of specialized tissue, differentiate into recognizable lens epithelium with the production of lens-specific proteins. These cells are regulated by physiologic mechanisms that maintain the status quo of the organism. If the optic cup does not interact with the overlying head ectoderm, head cells that would normally become lens would form skin or its derivatives. Whether differentiating into lens or into head ectoderm, mitosis is required for the differentiation to become complete, and the process is presumably controlled by the microenvironment of the cells.

Important gaps in our knowledge become obvious from the foregoing oversimplified sketch. In the first place, we do not know if differentiation requires an extracellular humoral stimulus, cell–cell contact. Possibly the mechanism for diversification is inherent in the cell and is merely expressed in favorable cellular environments. Although much is said about environmental signals and appropriate environments for differentiations, little or nothing is known about their nature. Although we recognize that potential is restricted as specialization occurs, little information is available on the mechanism of this restriction, including its initiation.

Much of what is known about these problems has been derived from amphibian systems that are easier to manipulate than mammalian systems. From these studies it has been found that signals from the cellular environment are required to direct the differentiation of cells and tissues in many animal species. Since their differentiation is regulated by the environment, the eggs of these organisms are called *regulative eggs*. The eggs of a contrasting group of organisms have cytoplasmic components that are heterogeneously arranged and that, at the time of mitosis, are assorted unequally so that the resultant blastomeres each receive different components. The blastomeres divide in accordance with their cytoplasmic inheritance and more or less indifferently to the cellular environment. These eggs are called *mosaic eggs*. The implication is that in some systems a mechanism may be built into the cell for sequential diversification by means of a cytoplasmic program that is relatively independent of specific external environmental signals. Other cells, however, may require specific environmental signals to differentiate. In either situation, development of the differentiated phenotvpe is dependent upon an appropriate cellular environment to support mitosis and to allow for the expression of the differentiation.

Spemann (1962) took cells from the dorsal lip of the blastopore of an amphibian at gastrulation and transplanted the cells to another gastrula; a second embryo developed at the site of implantation. This would appear to be an ideal situation in which to study the mechanisms of induction and the nature of the inducing stimulus. However, Holtfreter (1951) learned that induction is nonspecific. Frog dorsal lip will induce differentiation in other amphibians, and dead tissues and a host of chemicals are also active. None of these agents has any effect on cells determined for a particular differentiation, and the mechanism of induction is unknown. The only common denominator is the mitotically active competent responding cell in its overall environment; possibly the mechanism for diversification is built into a cell, as in mosaic eggs, and the intrinsic mechanism is triggered by nonspecific stimuli.

Study of inductive events in later development in mammals was undertaken by Grobstein (1953) to determine whether or not diffusible substances were involved. A responding primordial tissue, either salivary gland, kidney, or pancreatic primordia, was placed on one side of a millipore filter and an inducing tissue was placed on the other in a tissue culture dish. The formation of tubules in the primordial tissue indicated an induction.

Since cell-to-cell contact was presumably excluded in these experiments, tubulogenesis was believed to be mediated by one or more diffusible factors capable of stimulating DNA synthesis and morphogenesis. Ronzio and Rutter (1973) have isolated a substance, mesenchymal factor (MF), from mesenchyme that is capable of causing tubulogenesis of primordial pancreatic epithelium of the chick. MF is a noncollagenous protein that is not found in epithelium. It can be adsorbed to plastic beads and used to induce tubulogenesis of pancreatic epithelial primordia in the absence of mesenchyme. DNA, RNA, and protein syntheses, as well as DNA polymerase activity, are increased in the presence of this factor.

The situation is not without controversy. Saxen (1975) has examined millipore filters that separate epithelial primordia from mesenchyme and found mesenchymal processes penetrating the filter and forming a complex network on the epithelial side. No extracellular material was found in these processes and it was concluded that induction in this system depends upon cell-cell contact.

Steroids at later stages of development affect the differentiation of some tissues and not of others. Thus, as better and better systems of analysis are developed, specificities for inducing molecules, currently masked by the complexity of the systems, may be discovered.

In conclusion, proliferation and differentiation are orderly processes responsible for development in the embryo and tissue renewal in the adult. There are stringent rules governing these processes, and although the rules may differ in embryonic and adult states, the actual mechanisms appear to be similar wherever they occur, including neoplasms.

As for differentiation in neoplasms, studies on teratocarcinomas indicated the same orderly sequence of development that occurs in an embryo. Embryonal carcinoma differentiates a layer of endoderm, and presumably as a result of their interaction, mesoderm develops from embryonal carcinomas. Cartilage and muscle never develop directly from embryonal carcinoma; they only develop from the appropriate mesoderm. Similar lessons have been learned from the study of squamous cell carcinomas.

The mechanism of cellular differentiation is now believed to be the result of a precise and sequential read-off of selected parts of the cellular genome. In other words, the genome is not structurally altered during the process of differentiation; instead, its expression is controlled by repression or activation of the appropriate genes. The controlling molecular switches have been variously attributed to histones, nonhistone chromosomal proteins, and to other chemicals. As will be discussed in the section on nuclear transplantation, it would appear that the cytoplasm is critically important in controlling nuclear behavior and in controlling gene expression. We will refer to these observations and concepts when we discuss the heritable features of neoplasia.

Normal stem cells as the cells of origin

of tumors

From a theoretical standpoint, two bits of evidence point to stem cells as the precursor cells of epithelial tumors (Pierce, 1974). It is known that neoplasms occur only in tissues capable of mitosis and that the neoplasm will resemble its tissue of origin. If, for example, a tumor arises in skin painted with carcinogens, the tumor will be a skin tumor, not a tumor of brain and not a tumor of bone. Similarly, carcinogen applied to the bladder gives rise to a bladder tumor and not a lung tumor in the bladder. Virus-infected fibroblasts *in vitro* give rise to transformed cells which, when transplanted *in vitro*, develop into fibrosarcomas, not into adenocarcinomas. Thus, it can be deduced that the target cell in carcinogenesis is cycling and is already determined for a particular differentiation. The only cells that fulfill these definitional requirements are the stem cells of normal tissues and their partially differentiated progeny, which through controlled rates of proliferation and differentiation renew or maintain the mature tissue.

Direct evidence for the stem cell origin of a tumor is provided by the demonstration of Stevens (1967) that testicular teratocarcinomas of mice are derived from primordial germ cells. Since teratocarcinomas of strain 129 mice are often present at birth, Stevens (1962) examined fetal testes to determine when the tumors first appeared. The most rudimentary ones were observed in testes of 15-day-old fetal mice; by extrapolation Stevens calculated that the carcinogenic event must have occurred on or about the twelfth day of fetal life.

Stevens (1967) dissected genital ridges from mouse embryos of 12 days' gestational age and transplanted them into the testes of adult mice. Half of the transplants differentiated into ovaries and were discarded; the other half differentiated into fetal testes. Eighty percent of the fetal testes contained foci of embryonal carcinoma cells that were recognizable 7 days after transplantation. These foci quickly developed into classical teratocarcinomas. By incorporating a gene for the absence of germ cells into the genome of strain 129 animals, the 80% incidence of tumors in fetal testes was reduced to near 0%.

Light and ultrastructural studies of transplanted genital ridges undergoing carcinogenesis showed the close resemblance of embryonal carcinoma cells to the primordial germ cells from which they originated (Figs. 5–2 and 5–3). Both cell types were characterized by an exceedingly undifferentiated cytoplasm with a few mitochondria and sparse elements of rough endoplasmic reticulum. The tumor cells were rapidly proliferating and, as a consequence, the ribosomes were organized in polysomal configurations compatible with the pattern associated with synthesis of cell cytoplasm. The primordial germ cells were in a resting phase and their ribosomes were in a dispersed state (Pierce, Stevens, and Nakane, 1967). Similarly, Stevens (1968) and Damjanov *et al.* (1971) have produced teratomas by intratesticular grafts of fertilized ova. Fertilized ova, blastomeres, and primordial germ cells have equivalent degrees of differentiation.

Thus, for teratocarcinomas the stem cell of the species was identified as the cell of origin of the tumor. Although it may be argued that this cell is well differentiated for its particular purpose, its cytoplasm is undifferentiated in relation to that of a myocyte or a cartilage cell. In any event, after carcinogenesis, the normal stem cell gave rise to a malignant cell that closely resembled the normal cell in its degree of differentiation. The undifferentiated neoplastic stem cell, the embryonal carcinoma cell, proliferated and some of its progeny differentiated and ultimately produced tissues representing each of the three germ layers. The resultant teratocarcinoma is a caricature of embryogenesis.

If the stem cells of teratocarcinoma had their origin in normal undifferentiated germ cells, and closely resembled them ultrastructurally, it is conceivable that other neoplasms might also develop from undifferentiated cells. Franks and Wilson (1970) compared the ultrastructure of normal fibroblasts in tissue culture with fibroblasts transformed by oncogenic viruses. Normal and transformed fibroblasts had comparable degrees of differentiation; certainly the transformed and tumorigenic cells were no less differentiated than their normal counterparts.

It was decided to compare the differentiation of stem cells of breast

Fig. 5–2   A cytoplasmic bridge connects adjacent primordial germ cells (PGC). Note density of cell membrane of the bridge and character of cytoplasm of cells. Endoplasmic reticulum is less dense and polysomes are more numerous than in Sertoli cells (SC). Intercellular vacuoles (iv) between the Sertoli cell and the primordial germ cells presumably are a shrinkage artifact (11,000✕). From Pierce, Stevens, and Nakane (1967), courtesy of Journal National Cancer Institute.

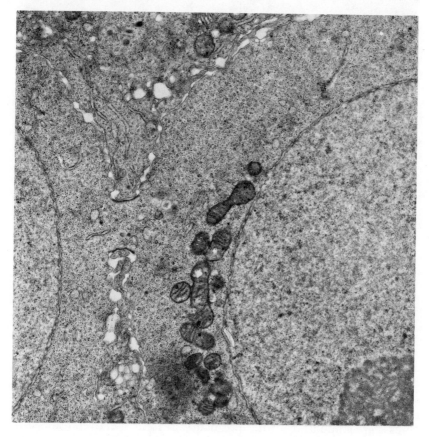

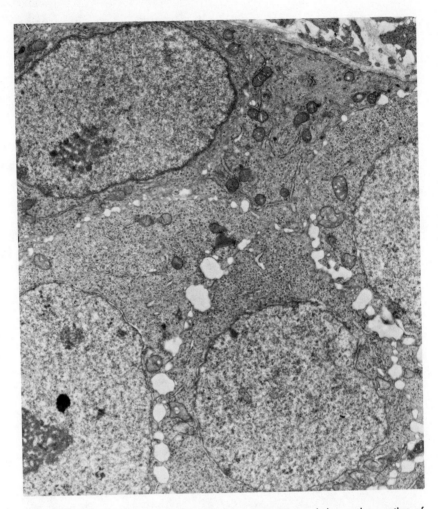

Fig. 5–3 Portions of three embryonal carcinoma cells (ECC) are below and a portion of the Sertoli cell (SC) and basement membrane (bm) of testicular tubule are above. The intercellular vacuoles (iv), presumably caused by a shrinkage artifact, are much more pronounced between the tumor cells than between the tumor cells and the Sertoli cells. Embryonal carcinoma cells are characterized by many polysomes, few profiles of endoplasmic reticulum (er), and widely scattered mitochondria (m). Sertoli cells in juxtaposition to tumor cells are characteristically well differentiated (9,500✕). From Pierce, Stevens, and Nakane (1967), courtesy of Journal National Cancer Institute.

and colon with that of their respective adenocarcinomas to determine if a pool of undifferentiated cells existed that could serve as targets in carcinogenesis (Pierce *et al.*, 1977). The simple ducts of the resting breast are stimulated during pregnancy to form the alveoli and ductules of the lactogenic tissue, so the stem cells of the breast must be contained in the ducts. The stem cells in the resting breast and their malignant counterpart in carcinomas of the breast were labeled with tritiated thymidine and their ultrastructures were compared by using autoradiography with the electron microscope (Pierce *et al.*, 1977). The normal and malignant stem cells were comparable in their degrees of differentiation. Each had sparse elements of rough endoplasmic reticulum and an atrophic Golgi apparatus. The polysomes were unattached to membranes, the configuration for synthesis of cell cytoplasm. There was no evidence of secretion.

Just as normal stem cells give rise to lactating cells, some of the progeny of stem cells of mammary adenocarcinomas differentiated into cells with secretion granules (see p. 44). These functional cells were postmitotic, indicating that normal and neoplastic stem cells have the same end point. The undifferentiated appearance of a cancer is the result of a predominance of stem cells and their partially differentiated progeny in relation to differentiated elements; it does not result from dedifferentiation. What is important from the standpoint of carcinogenesis is the realization that malignant stem cells can originate from undifferentiated normal cells, thereby obviating a need for the concept of dedifferentiation. It is unnecessary to postulate accidents of embryogenesis, sequestration of undifferentiated cells, or other highly unlikely explanations to account for the cell of origin of tumors.

Further support for the idea that normal stem cells are as undifferentiated as malignant stem cells was obtained in a comparison of stem cells of the colon and adenocarcinomas of the colon. Chang and Leblond (1971) identified two stem lines in the crypts of colonic epithelium: mucous cells and vacuolated cells; the latter differentiate into columnar cells. Using autoradiography with the electron microscope after injection of tritum-labeled thymidine, the vacuolated and mucous cells of normal colon proved to be no more differentiated than their counterparts in the adenocarcinoma (Pierce *et al.*, 1977).

Seilern-Aspang and Kratochwil (1962) injected benzo(a)-pyrene into the skin of newts and described the changes leading to carcinomas. The tumors each began in a separate mucous gland, and the earliest detected morphological change occurred in the germinative (stem cell) layer of the basal part of the glandular pockets. These cells commenced rapid division with the eventual development of metastasizing carcinomas. Further support of the idea that carcinogenesis involves undifferentiated cells is provided in the review by Siminovitch and Axelrad (1962).

The evidence is now overwhelming that normal stem cells, the reactive cells of tissue, are the target in carcinogenesis. Appreciation of this fact leads to an understanding of the relationship between benign and malignant tumors.

Origin and relationship of benign
and malignant tumors

The relationship of benign and malignant tumors has been unsatisfactorily explained in the past. Some believe that benign tumors are a stage in the development of malignant ones, an idea based on the observation that when chemical carcinogen is applied to the skin, many benign tumors appear first, followed by a few malignant ones. If the Shope papilloma virus is inoculated into the skin of a wild cottontail rabbit, warty benign tumors develop first, followed later by the develop-

Fig. 5–4 Model of cell renewal. It is postulated that a carcinogenetic insult could involve any of the cells capable of synthesizing DNA. Each would respond to the oncogenic stimulus by forming a stem line for a tumor. When initiated, the stem cells at the left would give rise to malignant stem cells that would morphologically resemble the normal counterpart and that, because of the extensive changes in the stem cell membranes, would appear foreign in their tissue of origin. A latent period of years would be required for them to achieve a critical number of cells capable of expressing their environment. The almost completely differentiated cell still capable of one round of mitosis when initiated would become the stem cell of the benign tumor. It would closely resemble its normal counterpart and would find all of the ingredients necessary for expression of its phenotype in the normal environment. Other cells would respond with varying degrees of malignancy and periods of latency. The tumor that would result initially would be dominated by the benign components capable of expressing their phenotype, but with time the more malignant elements would outgrow the benign ones. This process of change in properties of a tumor with time has been called *progression*.

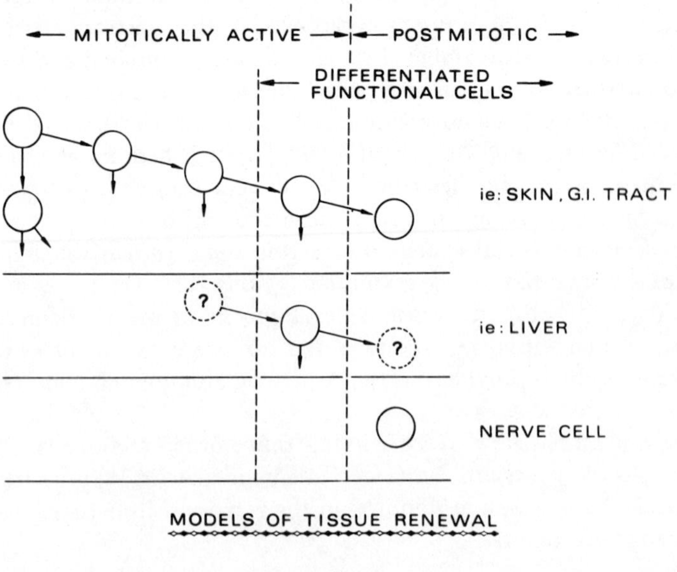

MODELS OF TISSUE RENEWAL

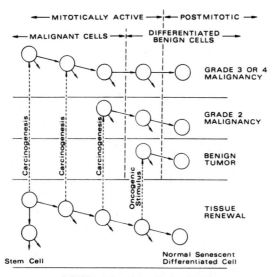

DIFFERENTIATION AND NEOPLASIA

Fig. 5–5 In this scheme the undifferentiated stem cells are at the left of the figure. They are determined for a differentiation, lack the molecular characteristic of it, and have the capacity for division only. At some point, progeny cells begin accumulating the molecules of differentiation in their cytoplasm. These cells are still capable of a round of mitotic activity and their progeny give rise to the differentiated functional senescent cell at the right of the figure. It is not known whether there are undifferentiated precursors of normal liver cells or whether liver cells ever become senescent. Nerve cells do not divide in the adult and there are no tumors of adult nerve cells.

ment of metastasizing carcinomas. When newborn mice are injected with polyoma virus, benign and malignant tumors develop. Individuals with multiple benign polyps of the colon invariably develop invasive adenocarcinoma of the colon, which again could be interpreted as indicating that benign tumors may develop into malignant ones. From these data it can be deduced that a carcinogenic stimulus can give rise to both benign and malignant lesions and that the benign tumors appear before the malignant ones, but there are no data that one gives rise to the other. The explanation for these phenomena probably is found in the reacting cells and their environments (Pierce, 1974). The state of differentiation of the reacting cells probably determines whether or not the resulting neoplasm will be benign or malignant; the environment is responsible for the appearance of benign tumors before malignant ones.

The process of tissue renewal is illustrated in Fig. 5–4 and the postulated effects of oncogenic stimuli are superimposed upon this scheme (Fig. 5–5). The determined and undifferentiated stem cells have but a single function: They can each divide into two cells. One of these cells undergoes maturational changes and divides again, and the cycles of maturation and division are repeated until renewal of functional post-mitotic cells has been accomplished.

The stem cells lack the differentiated molecules characteristic of the tissue, but they have the potential for this expression and pass it on to their progeny. The potential is fully realized in the postmitotic cells forming the terminal differentiation. If the undifferentiated stem cell is the target in carcinogenesis, it would be converted to a neoplastic stem cell, which would closely resemble its normal counterpart ultra-structurally, and form a malignant tumor, recognizable as a caricature of the tissue of origin, as described in Chapter Three. On the other hand, if carcinogenesis targeted the almost terminally differentiated cell still capable of synthesizing one more round of DNA, a benign tumor would result. The least differentiated cells of the benign tumor would closely resemble ultrastructurally the normal counterpart, and there would be no cells less differentiated than the normal counterpart. Tumors with intermediate degrees of differentiation between these extremes could be expected if intermediate cells were targeted.

This concept raises several interesting questions. Why should a car-cinogenic stimulus that reacts with an undifferentiated stem cell result in a phenotype characterized by rapid growth, invasion, nitrogen trap-ping, and metastasis, whereas the same agent reacting with an almost terminally differentiated cell causes, at most, an enhancement of growth rate and unresponsiveness to physiologic stimuli? Obviously, the answer to this question is unknown, but it probably resides in the stability of genomic controls mediated at the gene level and by chromo-somal structure in the terminally differentiated cell. The chromosomal structure of the differentiated cell might preclude accessibility to the genes of agents effective in altering its expression (see Chapter Eight).

The explanation for the earlier appearance of benign rather than malignant tumors probably has to do with the response of the trans-formed cell to its environment. Transformation of the undifferentiated stem cell results in such marked alterations that it is foreign in its tissue of origin and is a latent malignant stem cell unable to express the malignant phenotype until a threshold number accumulates. The experiments of Grobstein and Zwilling (1953) established the notion that a threshold number of cells was necessary for tissue genesis to occur *in vitro* (p. 18) and indicated that the threshold number of cells is that number that creates the optimal microenvironment for pheno-typic expression. Thus, the undifferentiated malignant stem cells would have a prolonged latent period, but when a critical number are pro-duced, they would express the malignant phenotype, invade, metastasize, and be largely refractory to environmental stimuli because of their own self-created environment. The cells of the benign tumor, however, are little different from normal cells, and their optimal metabolic requirements are found in the tissue of origin. Therefore, a much shorter latent period would be required for the development of the benign tumor.

We are now in a better position to understand the observations of Foulds (1956) who postulated that tumors undergo progression, *i.e.*, the loss or gain of unit characteristics with time. If at the time of exposure to carcinogen a spectrum of cells in the normal renewal process are transformed, the initial phenotype of the tumor will be dominated by less malignant cells because they are "less foreign" and proliferate maximally. As more malignant cells reach threshold number and express their phenotype, the nature of the tumor would change in favor of invasion, metastasis, and lack of differentiation. The observations of Greene (1951) and Foulds (1956) appear more and more to be selections of cells best able to survive under the conditions of growth rather than progressive alteration of the cells themselves. Yet one cannot ignore the nagging thought that the genomic instability of cancer cells provides the diversity on which selection can act to produce more malignant cells.

SIX

Carcinogenesis

Introduction

In addition to identifying and eliminating carcinogenic hazards, the ultimate goal of studies of carcinogenesis is to understand the intracellular mechanisms whereby the normal stem cell is converted to a malignant stem cell. Several hundred chemicals, both RNA and DNA viruses, and physical agents such as x-ray and ultraviolet light are capable of causing cancer. Although there is great diversity in the causative agents, the end point of the cellular reaction is always the same: the development of neoplastic stem cells and, thereby, the neoplasm.

It is not known whether or not chemicals, viruses, and physical agents operate through the same intracellular mechanism. This is a tempting possibility because of the limited number of ways in which cells can react to physiologic or injurious stimuli. It could also be postulated that there are multiple mechanisms by which the effects of various carcinogenic agents are mediated.

Determination of what is carcinogenic has been unexpectedly complex. For example, a chemical that is carcinogenic for one species might not be for another. The prolonged latent periods between application of the agent and development of the tumor make testing not only laborious, but also impossible for the elucidation of intracellular mechanisms of carcinogenesis. Similar problems have been encountered by viral oncologists: A virus that produced a neoplasm in one species might produce an inflammatory disease in another (Duran-

Reynals, 1941). For example, the adenovirus that causes upper respiratory disease in man transforms hamster cells. Not all tissues in susceptible animals become transformed and there is a long latent period between the time of exposure to virus and the development of tumors.

In an effort to exclude host influences on the carcinogenic process, Earle (1943) studied cells in culture. Before these experiments were completed, it became apparent that normal cells grown for prolonged periods as either explants or monolayers spontaneously transformed to what appeared to be a neoplastic state. This compounded the problems of studying carcinogenesis in tissue culture, but methods have now been evolved that allow for quantitative studies to be performed. Although tissue culture appears to be a most promising tool for the study of carcinogenesis, it is still not the ultimate in test systems. An agent that causes tumors when administered to an animal does not necessarily transform cells explanted from the same animal because metabolic conversion by the intact animal may be required to convert the agent to an active carcinogenic form. Most of the neoplastic transformations *in vitro* convert mesenchymal cells to sarcomas; however, the most common malignancies in mammals are carcinomas, not sarcomas. Even so, if the fundamental mechanism by which transformation *in vitro* takes place is learned, it can be assumed that we will have gone a long way in understanding neoplastic conversion.

Discussion will now center around chemical carcinogenesis, viral carcinogenesis *in vivo*, and finally, transformation.

Chemical carcinogenesis

Chemical carcinogenesis had its clinical beginning with the observation of Potts in the late eighteenth century that chimney sweeps had a high incidence of scrotal cancer, which was related to soot that was difficult to remove from the genital areas. This observation led to public health legislation designed to protect chimney sweeps from this occupational hazard. Thus began the combined effort of science and government to identify and eliminate carcinogens in industry and in the enviroment (Hueper, Wiley, and Wolfe, 1938). Table 6–1 lists some of the agents that have been implicated in the etiology of human neoplasms. Before environmental carcinogenesis can be appreciated it is necessary to understand chemical carcinogenesis and transformation.

TABLE 6–1

Chemicals Carcinogenic in Humans	Target Tissues
Soots, tars, oils	Skin, lungs
Cigarette smoke	Lungs
2-Naphthylamine	Urinary bladder
4-Aminobiphenyl	Urinary bladder
Benzidine	Urinary bladder
N,N,bis(2-chloroethyl)-2-naphthylamine	Urinary bladder
Bis-(2-chloroethyl)-sulfide	Lungs
Nickel compounds	Lungs, nasal cavity
Chromium compounds	Lungs
Asbestos	Lungs, pleura, ? others
Cadmium oxide	Prostate
Diethylstilbestrol (DES)	Vagina
Aflatoxin	Liver
Arsenic	Lungs, skin
Auramine	Urinary bladder
Vinylchloride (VCM)	Liver
Thorotrast	Liver

SOURCE: Taken from J. A. Miller, *Cancer Research*, 30 (1970), pp. 559–576. See also E. Boyland, *Progress in Experimental Tumor Research*, 11 (1969), pp. 222–234.
NOTE: This is a partial list.

Modern experiments in carcinogenesis began with the work of Yamagiwa and Ichikawa between 1915 and 1917. The application of coal tar to the ears of 137 rabbits two or more times per week resulted in tumors in most of the 22 rabbits still alive after 150 days. After 360 days seven of these were carcinomas, two with lymph node metastasis. Soots and tars were subsequently analyzed chemically to identify the active substances, and in 1930 Kennaway (1955) successfully isolated 3,4-benz(a) pyrene, a potent carcinogen. He subsequently postulated that 1:2:5:6-dibenzanthracene would be a carcinogen on the basis of its structure, and he then proved it to be one. Now hundreds of chemical carcinogens with extremely varied structure have been identified and classified, and their toxicity, carcinogenicity, and species specificities have been described.

There have been two approaches to examining the mode of action of carcinogens. In one approach, cancers were studied biochemically in an effort to determine variations from normal that would pinpoint essential neoplastic attributes, indicate how carcinogens work, and suggest means of therapy. Although this approach has resulted in many important discoveries, it has not led to an understanding of carcinogenesis (see Chapter Five). It is as difficult for biochemists to describe the mechanism and origin of tumors from analysis of cancers as it would be to expect them to determine (by biochemical analysis of an adult) the mechanism whereby the adult developed from an ovum. Like development, carcinogenesis is a process, and it is necessary to study the process on a time-controlled basis rather than just the end result.

The other approach to the understanding of chemical carcinogenesis has been to study the metabolism of chemical carcinogens thereby hoping to elucidate the biochemical events which, in summation, produce the neoplastic phenotype. This approach has been fruitful, although it is still not known whether the mechanism of carcinogenesis is genetic or epigenetic.

It will serve no useful purpose to review in detail these very complex chemical experiments; suffice it to say that chemical carcinogens can be divided into two categories. One is exceedingly *reactive* and capable of causing neoplastic conversion at the site of application. Such *ultimate* carcinogens cause squamous cell carcinoma if applied to the skin or fibrosarcomas if injected into the dermis. Chemicals in the other category often do not cause tumors at the site of application; the tumors develop in other parts of the body. This led Miller and Miller (1971) to postulate that some agents might be precarcinogenic or very weak molecules that require metabolic conversion to an active form by the host. The enzymes responsible for the conversion were found in the microsomal fraction of those cells that often becomes neoplastic (Gelboin, Kinoshita, and Wiebel, 1972). In the case of the liver, the enzymes are part of the mechanism of the detoxification of poisons. Normally, liver enzymes render poisons nontoxic, soluble, and available for excretion in either bile or urine.

2-acetylamino-fluorene (AAF) causes tumors in many organs and in many species: Rats and mice develop liver, ear duct, breast, lung, and bladder tumors and leukemia. The metabolism of AAF has been studied by Miller (1970) and illustrates how the detoxification of poisons by the liver can give rise to a carcinogen. When the ring carbons of AAF are hydroxylated by microsomal enzymes and the resulting alcohols are esterified, a soluble noncarcinogenic glucuronide is produced that is excreted in the urine. However, if AAF is hydroxylated in the amide portion of the molecule by the microsomal oxidases, a moderately carcinogenic molecule, N-hydroxy-acetylaminofluorene (N-OH-AAF), results.

It turned out that guinea pigs lack the enzymes for converting AAF to N-OH-AAF; therefore, they do not develop tumors when they are fed AAF, but they do when they are fed N-OH-AAF. This demonstrates that the species variation in response to carcinogens is, in part, dependent upon the ability of the species to convert a precarcinogenic molecule or a weakly carcinogenic molecule to a more active form. When N-OH-AAF was synthetically converted to the ester, N-acetoxy-AAF, the latter proved to be so carcinogenic that it is considered to be the ultimate carcinogen. *In vivo*, N-OH-AAF is probably sulfonated by soluble sulfotransferases in the liver, and the sulfonic acid ester of AAF is believed to be the natural ultimate carcinogen.

Just as the guinea pig was refractory to the effects of AAF because it lacked a particular enzyme, the male rat is more susceptible to the hepatocarcinogenicity of N-OH-AAF because it has a higher N-OH-AAF sulfotransferase activity than the female rat. This explains one sex difference in carcinogenesis and suggests a possible effect of hormones on the enzymes necessary for the conversion of carcinogens to their ultimate reactive form. It also suggests a reason for tissue specificity of certain carcinogens.

The route of excretion of the activated molecule is also important in determining the tissue specificity of carcinogens. For example, 2-naphthylamine is a cause of cancer of the human urinary bladder (Hueper, Wiley, and Wolfe, 1938). This compound is converted to hydroxy-2-naphthylamine in the liver, excreted in the urine, and stored in the bladder. Individuals developed bladder tumors, not because the precarcinogen was converted to carcinogen in bladder mucosa, but because the ultimate carcinogenic molecule was excreted in the urine.

Many other carcinogens are converted from inactive to active forms. Cycasin, a soluble carcinogen found in cycad nuts, is hydrolyzed by bacterial β-glucosidase to an active carcinogen in the gastrointestinal tract where it produces tumors in animals. The metabolite can alkylate nucleic acids *in vivo* and *in vitro* (Shank and Magee, 1967). The liver carcinogen, ethionine, an analog of methionine (Farber, 1973 a and b; Sarma, Rajalakshmi, and Farber, 1975), donates ethyl instead of methyl groups to RNA, and to a lesser extent, to DNA. Why ethylation of RNA or DNA should cause cancer, if in fact it does, is not known.

Identification and quantitation of the enzymes responsible for metabolic conversion of chemical carcinogens may have important clinical implications. For example, polycyclic aromatic hydrocarbons, potent carcinogens in experimental animals, are metabolized to active forms by certain oxidases. One of these, aryl hydrocarbon hydroxylase (AHH), is present in a variety of tissues. AHH is inducible by a variety of agents, and in the mouse this inducibility is determined by a single autosomal gene. Kouri, Ratrie, and Whitmire (1973) have found a close correlation between susceptibility to methylcholanthrene-induced subcutaneous tumors in mice and the gentically determined inducibility of AHH. Kellerman, Luyten-Kellerman, and Shaw (1973), studying AHH inducibility in cultured human leukocytes, suggest that the American population can be divided into high, intermediate, and low inducibility groups. This variation is due to two alleles at a single locus. An association has been found between high AHH inducibility and bronchogenic carcinoma in smokers; no such association could be demonstrated in patients with other forms of cancer (Kellerman, Shaw, and Luyten-Kellerman, 1973). Although the relationships cited above require further documentation and clarification, the implications with regard to genetically determined differences in susceptibility to potentially carcinogenic agents are intriguing. The extension of this principle to other carcinogens, enzymes, and tumors might allow the identification of individuals at high risk of developing cancer when exposed to particular agents.

In summary, some agents in their native form cause cancer, and others require metabolic conversion to become active carcinogens. Thus, there are several factors that determine the susceptibility and tissue incidence of tumors; these include species-specific enzymes, tissue-specific enzymes, hormonal influences on enzyme levels, and the mode of uptake or elimination of the carcinogen. This is the basis for studies to control environmental carcinogens (see p. 109).

Viral carcinogenesis

Early experiments in which leukemia and sarcoma of chickens were transmitted by cell-free filtrates were not accepted as proof of a viral etiology for tumors. In 1908 Ellerman and Bang succeeded in transmitting a leukemia to normal chickens by using a cell-free extract. Two years later Rous (1910, 1911) discovered a sarcoma in a hen that could be maintained as a transplantable tumor and that eventually was *passaged* with cell-free filtrates.

These observations were confirmed and extended to other tumors of birds a few years later, but they were far ahead of their time. Since leukemias were not considered neoplastic diseases by many investigators, the observations were discounted. The inflammatory reaction that often accompanied the sarcomas of chickens led others to believe that this was a peculiar kind of proliferative inflammatory response and not necessarily the equivalent of neoplastic disease in mammals. Finally, it was impossible to produce tumors in mammals with cell-free extracts, and the issue of whether or not tumors were caused by viruses was considered closed by many.

In 1932 and 1933 Shope reported the occurrence of fibromas and papillomas in wild cottontail rabbits that were transmissible by cell-free extracts. Within a month after inoculation of a cell-free extract of a papilloma, a warty pigmented tumor developed that seldom regressed. Metastasizing carcinomas developed in some of the animals months later. Whereas the virus was readily identified in papillomas by infectivity tests (injection of cell-free extracts of tumors into a rabbit), results of infectivity tests made upon the carcinomas were invariably negative. When passaged in domestic rabbits, the papilloma virus induced papillomas and carcinomas. Infectivity tests performed on the carcinomas were negative (Shope, 1937).

It is to be noted that the papilloma virus discovered by Shope was the first demonstration that metastasizing carcinomas could be produced by cell-free extracts of naturally occurring mammalian tumors. Cell-free extracts of fibromas of cottontail rabbits (Shope, 1932) produced fibromas that regressed after a number of months, but when the virus was injected into young domestic rabbits, it produced metastasizing neoplasms.

Concomitant with these studies were others in which inbred strains of mice were produced that had an exceedingly high incidence of adenocarcinomas of the breast. These observations were interpreted as supporting the notion that carcinoma was a genetic disease. Kortewig (quoted by Bittner, 1947) noticed that if high and low tumor strain animals were crossed, the F-1 females had an incidence of adenocarcinomas of the breast that approximated that of the mother. If the mother was from a high tumor strain, the F-1 females had a high incidence of tumors, and conversely, if the mother was from a low tumor strain, the F-1 females had a low incidence of breast cancer. This suggested the presence of an extra-chromosomal factor, an idea confirmed when Bittner (1939) demonstrated, through foster nursing experiments, that there was a factor in the milk of high tumor strain females that was responsible for the high incidence of tumors in the F-1 generations. Although animals received the factor shortly after birth, the tumors did not develop until they were at least 40 weeks of age and their production was modulated by the genetics of the host (Heston, 1965) as well as by endocrine background.

Because the viral etiology of cancer was in disrepute at the time, Bittner (1947) referred to this agent as the milk factor. The significance of the notion that animals must receive the factor (virus) immediately after birth for production of tumors was not capitalized upon until Gross (1951) successfully passed leukemias of AK mice by injection of cell-free extracts into newborn C_3H mice. Specifically, 7 of 14 animals so inoculated developed leukemia 8 to 11 months later, and 4 of 6 C_3H mice injected with an ultrafiltrate of AK embryos developed leukemia about 8 months later. Animals injected with these filtrates more than 12 hours after birth did not develop leukemia. The results of these experiments were considered highly controversial because of the low incidence of leukemia obtained. In 1953 Gross attempted to improve upon the incidence of disease by changing the mode of preparation and made another important discovery. Although only 9 of 84 animals developed leukemia 8 months after injection of ultrafiltrate, 15 of the 84 developed undifferentiated adenocarcinomas of salivary glands by 3 months of age. None of the latter animals had leukemia. Only one of 54 control animals developed a salivary gland tumor.

Gross postulated that leukemia was transmitted from parent to offspring (vertical transmission) and that the viral suspension obtained from AK mice was probably made up of two viruses, one that produced leukemia and one that produced solid tumors. Stewart (1953) made comparable observations independently, but he found it difficult to demonstrate the virus in the parotid tumors. In an effort to increase the yield, Eddy, Stewart, and Berkeley (1958) and Eddy *et al.* (1961) grew the virus in cultures of embryonic mouse cells. Filterable extracts of these culture fluids gave rise to tumors in more than 75% of recipients. Tumors developed not only in the parotid gland, but also in the adrenal and many other organs. This virus did not produce leukemia. Thus, two viruses were present in the original suspension; the one producing leukemia was not cytopathic in tissue cultures, whereas the other, which gave rise to solid tumors when injected into mice, caused rapid destruction of embryonic mouse cells in tissue culture. The cytopathic virus was named *polyoma* because it gave rise to multiple kinds of tumors in mice and tumors in multiple species (Stewart *et al.*, 1957; Eddy, and Borgese, 1958). This was reassuring to oncologists who couldn't envision a different virus for each of the many kinds of tumors known. Many of the tumors produced by polyoma were benign, but others were highly malignant metastasizing sarcomas. Epithelial tumors were also produced (Defendi, Lehman, and Kraemer, 1963).

In early 1961 Eddy *et al.* isolated a virus from tissue cultures of Rhesus monkey kidney cells that caused vacuolization of these cells and produced tumors when injected into newborn hamsters. This was named Simian virus 40 (SV_{40}) (Sweet and Hilleman, 1960). Although it caused cytopathic effects and replicated in monkey kidney cultures, it transformed mouse cells and caused tumors when injected in newborn hamsters. Adult mice, rabbits, and guinea pigs were not susceptible, probably because of immunologic resistance. The hamster is notable for its inability to mount effective immune responses and is the animal of choice for viral oncologic studies.

These were most important discoveries because they showed that a virus might cause at best mild degenerative changes in cells of one species and tumors in another. In this regard, Duran-Reynals (1941) showed that under certain circumstances Rous sarcoma virus produced only an inflammatory reaction. Moreover, SV_{40} was not recoverable from tissue culture supernatants of transformed cells. Polyoma was recoverable from tissue cultures of mouse cells (the lytic system), but infected hamster cells became transformed and did not produce virus.

Trentin, Yabe, and Taylor (1962) demonstrated that certain adenoviruses, which are commonly found in the upper respiratory tract of human beings, could cause tumors in hamsters. Specifically, type 12 and subsequently types 18 and 31, were found to produce rapidly growing sarcomas at the site of inoculation when injected into hamsters. It is now known that these viruses will also produce tumors in rats and certain strains of mice. It is unknown whether or not they cause cancer in human beings. More recently, a virus has been found that induces leukemia in cats and that is transmissible to cells of a variety of species, including man.

Many attempts have been made to demonstrate a viral etiology of human tumors. Ethics preclude the inoculation of cell-free extracts of human tumors into newborn babies as had been done in experimental animals. Even though there is strong circumstantial evidence that viruses are implicated in the etiology of some human tumors, it has been impossible to fulfill Koch's postulate.* In 1919 Wile and Kingery successfully transmitted human warts with cell-free extracts. Since these were benign tumors that spontaneously regressed, little attention was paid to the experiments. In 1958 Burkitt described a type of lymphoma, endemic in certain areas of Africa, that is often localized in the jaws of children and eventually causes death. Although viral particles are not detectable in these cells *in vivo*, Epstein, Achong, and Barr (1964) discovered herpes-like particles (the virus is known as Epstein-Barr virus or EBV) in cell cultures of these tumors. EBV antigen was then demonstrated in the cells of Burkitt's lymphoma, and EBV-DNA has been demonstrated by molecular hybridization (Zur Hausen and Schulte-Holthausen, 1970). Cell homogenates prepared against EBV-infected cells are capable of transforming normal lymphocytes. Patients with the disease have high titers of anti-EBV.

There is an interesting relationship between EBV and infectious mononucleosis. Virus is recovered from throat washings and the patients have antibodies to EBV. Why some people infected with EBV

---

* Koch's postulate: see footnote, p. 57.

get a self-limited infectious disease and others a lymphoma is not known.

Thus, the inferences are strong that EBV causes a human lymphoma. A similar DNA virus has been shown to cause neoplasms in monkeys; here, at least, Koch's postulate has been fulfilled. Normally, herpes virus saimiri replicates in squirrel monkeys but causes no disease. However, if injected into marmoset monkeys, lymphomas develop (Wolfe, Falk, and Deinhardt, 1971). Infectious virus is not produced *in vivo*, but when the cells are plated *in vitro*, viral replication takes place with release of infectious virus. When these particles are reintroduced into marmoset monkeys, tumors develop.

Patients with cancer of the cervix have high titers of antibody to herpes simplex virus-2 (HSV-2). Viral particles have been visualized in degenerating cervical carcinoma cells *in vitro* (Aurelian *et al.*, 1971). Similar viruses have been observed in human prostatic carcinoma cells, raising the possibility that cancer of these organs may be viral in origin and venereal in transmission.

Because C-type (oncorna) viruses have been found in chickens, rodents, cats, and primates, and these viruses can produce sarcomas or leukemias in experimental animals, it has been postulated that they might cause tumors in man. C-type viruses have been observed in electron micrographs of human tumors and cell lines derived from human tumors (see section on electron microscopy of RNA tumor viruses, p. 103). A C-type virus isolated from human sarcoma cells was tumorigenic in cats. This virus, isolated by McAllister *et al.* (1972), and called RD-114, appears to be a contaminating feline virus that can grow in human cells. Studies have shown that much of the genetic information in the virus is homologous to feline DNA. Gallagher and Gallo (1975) have isolated an RNA tumor virus from the cells of a patient with chronic myelogenous leukemia. Although this virus has been reisolated several times from the cells of this patient, it has not been isolated from the leukocytes of other leukemic patients, and its RNA is not homologous to human DNA. The antigenic characteristics of the viral proteins and studies of nucleic acid homologies suggest that this virus is related to a virus isolated from apes.

Another approach to the detection of viral genome in mammalian cells has been the use of radioactive DNA probes. These probes are synthesized by using an RNA dependent DNA polymerase (reverse transcriptase) from either animal virus templates or a template indigenous to the tumor cells. Successful molecular hybridization of the labeled nucleic acids to DNA from the tumor cells is indicative of viral genome. Spiegelman *et al.* (1972) believe that tumor virus RNA is present in human leukemic cells, in carcinomas of the breast, and in lung, gastrointestinal tract, and brain tumors. Similar studies on identical twins, one of whom had leukemia, suggested that there were integrated proviral sequences present only in the leukemic twin (Baxt *et al.*, 1973).

All of these and other evidences of a viral etiology for human cancer are tantalizing, but since none has demonstrated that there is a virus isolatable from human tumors that can cause a neoplasm in human beings, it is still uncertain whether or not these viruses are intimately involved in oncogenesis or are passengers in neoplastic cells.

Transformation

Transformation is believed by many to be the equivalent in tissue culture of carcinogenesis *in vivo*. Everyone recognizes that changes occur in normal cells in culture that are stable and heritable, and that closely mimic the characteristics of cells explanted from sarcomas.

Transformation can be spontaneous or induced by chemical carcinogens or oncogenic viruses and the latent period that has prevented sequential molecular analysis of carcinogenesis *in vivo* is reduced to a manageable length of time. Transformation would thus appear to be a valuable tool, but many oncologists question whether or not transformed cells are malignant and therefore useful in oncology. For example, transformed cells transplanted into suitable hosts *in vivo* may not produce tumors. Although the question cannot be answered at this time, when we understand transformation, we will require little further information to understand differentiation in neoplastic or normal cells.

Whereas pathologists (years ago) found that there is no one microscopic feature diagnostic of malignancy, cell biologists are still wrestling with defining criteria for transformation. An example of the problem is found in the experiments of Barski and Cassingena (1963) in which cell strains were developed from the lungs of mice. They were maintained as monolayers and two transformed lines were obtained—their transformation was demonstrated by increased growth rate, an indefinite life span, morphologic changes, and the appearance of an abnormal chromosomal pattern. Despite their identical features, only one of the two transformed lines formed tumors in animals. In this, and in many other instances, none of the features of transformation has been shown to correlate absolutely with tumorigenicity *in vivo*. Many transformed cell lines are capable of growth in soft agar, and there seemed to be a good correlation between growth in soft agar and tumor formation *in vivo* (MacPherson and Montagnier, 1964). Since many primary tumors will not grow in soft agar, this criterion is not absolute for malignancy.

Transformation has been studied intensively to determine whether or not it really is the *in vitro* equivalent of carcinogenesis *in vivo* and if transformed cells are malignant. The problem is exceedingly complex: It was pointed out in Chapter One that malignant tumors differ from benign tumors by intrinsic properties that will ultimately lead to death of the untreated host. Malignancy is a clinical concept. The argument has been presented that benign and malignant tumors arise by similar mechanisms. For instance, polyoma virus injected into a newborn mouse gives rise to benign and malignant tumors. Chemical carcinogens painted on the ears of rabbits give rise to benign and malignant tumors. Unless each of the target cells was identical in degree of differentiation, should not transformation, if similar to oncogenesis, involve a spectrum of cell lines, some the equivalent of benign tumors and the others of malignant ones? The former would not be tumorigenic, but the latter would produce tumors on transplantation, as is customary for malignant tumors. Irrespective of tumorigenicity, we would suggest that the process of transformation is the same in the benign and malignant situations. This is an idea that has escaped attention. Tissue culturists still search for the aspect of transformed cells that will correlate with tumorigenicity.

The dilemma of the oncologists in evaluating transformation and correlating it with *in vivo* malignancy is as real and difficult as the problems facing the pathologist attempting to make a diagnosis of malignancy based on the histologic appearance of a tumor. If transformation is, indeed, the *in vitro* equivalent of carcinogenesis, then its advantages as an investigative tool are obvious—the system would allow approaches precluded by the complexities of *in vivo* systems. The key word, however, is *equivalent*, and this nagging question necessarily leads to concern about the interpretation of observations based on transformation studies and their application to neoplasia. Transformation, whether spontaneous or induced, involves fibroblasts and not epithelial cells; yet epithelial cells give rise to the vast majority of mammalian tumors. The criteria used to score transformation and to describe its phenotype *in vitro* may not have relevance to epithelial tissues. In fact, the transformation of epithelial cells in culture by chemical carcinogens has been accomplished only a few times (Weinstein *et al.*, 1975) and cannot be effected in a reproducibly quantitative manner. Since epithelial cells are notoriously difficult to grow *in vitro*, this whole problem may simply be very difficult technologically.

Investigators must recognize that one of the limitations in using cell cultures as a model for carcinogenesis is that they deal with selected cells. When explants are taken and the cells are dissociated and plated as a monolayer, there is immediate and continued selection for those cells best able to survive under the conditions of the *in vitro* environment. Selection is even more extreme when cells are frozen for prolonged storage. Most tissue culture conditions are relatively anaerobic and those cells able to proliferate fastest under these conditions are selected. It is small wonder that certain mouse cultures, when serially propagated, consist almost exclusively of aneuploid cells with an indefinite life span after relatively few passages. These are characteristics frequently equated with transformation. Some investigators use such cells for studies of chemical or viral transformation. They score changes in morphology, growth in agar, or tumorigenicity as criteria of transformation, but they must accept that the cells are abnormal at the start and that they are probably studying only one phase of a multistep progression.

Finally, the tissue relationships and hormonal and nutritional factors necessary for normal development *in vivo* are lost in cell cultures. This loss may be more than offset by the ability of the investigator to control the *in vitro* environment, but, nevertheless, these factors may play important roles in carcinogenesis, and their absence may have important implications for transformation, particularly of epithelial cells.

In spite of all arguments, the transformed state can be induced by specific agents—quickly, reproducibly, and with a recognizable end point. The system is amenable to molecular analysis and when we understand the molecular basis for transformation, we will have learned much about differentiation and carcinogenesis.

Spontaneous transformation

Spontaneous transformation was discovered during studies of chemical carcinogenesis *in vitro*. Earle and Voegtlin (1940) and Earle and Nettleship (1943) attempted to duplicate, *in vitro*, the anaplastic changes observed in tumors *in vivo* and exposed cells of the mouse to chemical carcinogens in tissue culture. Despite the marked toxicity of even low concentrations of methylcholanthrene, strains of cells were developed that could be serially propagated and studied. Although carcinogen-treated cells were markedly altered, similar morphologic changes were observed in untreated cells maintained for at least 200 days *in vitro*. It must be stressed that they were never treated with carcinogen; yet when they were transplanted into isogenic mice, they gave rise to rapidly growing, transplantable, and metastasizing sarcomas indistinguishable from those obtained from the carcinogen-treated cultures.

Similarly, Gey (1947) described morphological changes in serially propagated rat fibroblasts, and subsequently Gey *et al.* (1949) obtained a progressively growing metastasizing sarcoma when cultures of these fibroblasts, maintained for long periods of time, were transplanted into suitable rats. Like Earle, Gey was positive that his cells had not been inadvertently exposed to carcinogen and attributed the neoplastic change to heterologous proteins in the tissue culture media. Much later, transformation was shown to occur in the absence of such proteins. In an effort to determine if transformation might have resulted from trace contamination by chemical carcinogens, Sanford *et al.* (1950) briefly exposed cultures of mouse fibroblasts to low levels of methylcholanthrene and then maintained them *in vitro* for prolonged periods. No change in incidence of sarcoma over control values occurred when these cells were transplanted. Later, other workers demonstrated that spontaneous transformation could take place in epithelial cells as well, but even today the morphological criteria for transformation of epithelial cells are not well characterized.

Through examining the circumstances of spontaneous transformation, it was seen that explanted fibroblasts usually grow luxuriantly for a variable period of time—about 60 days. Then they enter a dormant phase during which they utilize medium at progressively slower rates. In succeeding months most of them die. This is the phase of *crisis* and Hayflick and Moorhead (1961) decided that normal cells have an intracellularly controlled, finite life span. If, however, transformation occurs during the crisis, the pH of the medium changes dramatically and the cultures, now consisting of a new population of cells, grow luxuriantly. These newly transformed cells apparently originate from a few cells and not by conditioning of the culture as a whole.

Spontaneously transformed cells have distinguishing characteristics. They have an exceedingly uniform morphology and high nucleocytoplasmic ratios. They grow in multilayered tangles, uninhibited by contact, rather than in well-ordered arrays of flattened cells, and they quickly outgrow normal cells; instead of a finite life span, they are now capable of indefinite growth *in vitro*. Unlike the normal cells, they can be cloned easily and proliferate readily in low concentrations of serum. Many transformed cells grow as colonies in semisolid media, are heteroploid, and are agglutinable with plant lectins.

Their properties closely resemble those of tissue cultures derived from sarcomas, an observation that had led to the notion that they are, in fact, malignant and that the transformed state is the equivalent of malignancy. Transformed cells transplanted into animals usually grow quickly into sarcomas, strengthening the notion that transformation may be the equivalent of carcinogenesis.

Transformation with chemicals

Lasnitzski (1958) studied the effects of chemical carcinogens on organ cultures of mouse prostate. Doses of carcinogen that were toxic to the connective tissue elements resulted in epithelial hyperplasia with irregular crowding of cells, nuclear enlargement, and polyploidy. It was not until the 1960's that monolayer culture systems were devised in which effects of specific carcinogens could be distinguished from nonspecific ones. Probably the first successful experiments were those of Berwald and Sachs (1965) who observed foci of transformed cells after addition of methylcholanthrene or 3,4-benz(a)pyrene to monolayer cultures of mouse or hamster cells. Transformed cells had morphological alterations similar to those observed in cultures of *in vivo* tumors. They formed random patterns, grew indefinitely, and produced tumors on transplantation. When these cells were compared to normal cells, they were relatively resistant to the toxic effects of the carcinogen. From 10% to 25% of the cells were transformed and there was a lower percentage of transformed cells in crowded cultures, a feature these cells shared with those transformed by polyoma virus (Stoker and MacPherson, 1961). This might suggest that cell division that was not occurring under crowded conditions was important for transformation. After induction, cell division is necessary to fix the transformed state (Kakunaga, 1975).

Heidelberger (1970, 1973) studied monolayers of aneuploid fibroblasts derived from the prostates of C_3H mice. These cells grew indefinitely, were aneuploid, required high serum concentrations for growth, and did not produce tumors on transplantation even with a large inoculum of cells. Thus, they were abnormal, but they were not considered malignant. Chen and Heidelberger (1969) and Heidelberger (1973) exposed their aneuploid cell lines to methylcholanthrene for less than 24 hours. Whereas the original cells didn't produce tumors when 10^6 cells were injected into radiated C_3H mice, as few as 10^3 cells of the treated cells produced a tumor on transplantation. Single cells were exposed to methylcholanthrene. All surviving clones were transformed and gave rise to tumors when transplanted into mice. Mondal and Heidelberger (1970) found that toxicity of the carcinogenic agent was not related to the ability to transform cells, and they decided that ultimately all of the cells in the long-term cultures of methylcholanthrene-treated cultures were transformed but that the transformed state was expressed at different times.

Similar observations were made by DiPaolo and Donovan (1967). Primary cultures of Syrian hamster cells were exposed to benz(a)pyrene or methylcholanthrene. Toxicity was evidenced by the presence of granular cells, giant cells, ghost cells, and the detachment of cells from the substrate. These changes continued for approximately 30 days after exposure. Two and one-half months after exposure there was an increase in growth rate, decreased cell cycle time (from 24 hours to between 15 and 17 hours), and multilayering of cells. Interestingly enough, although the cells lacked the morphologic features of transformed cells, they had an indefinite life span; control cultures not treated with carcinogen had limited life spans (Hayflick and Moorhead, 1961). The transformed cells grew as colonies in soft agar and large numbers of cells were required to produce tumors in irradiated hosts. After 90 days the treated cultures were reexposed to carcinogen with little or no toxic effect. Aneuploidy as evidenced by the presence of one or two large submetacentric chromosomes was present. It is of interest that there was little or no pleomorphism in these cultures [unlike those of Heidelberger (1973), and Berwald and Sachs (1965)], yet these cells were tumorigenic when transplanted.

In later studies (DiPaolo, 1974), Syrian hamster embryo cells seeded at low densities on feeder layers of x-irradiated rat cells and treated with a variety of chemical carcinogens produced piled-up colonies of tumorigenic spindle cells within 9 days of treatment. As yet, diploid human cells have not been transformed by chemical carcinogens *in vitro*.

Transformation by viruses

When polyoma virus is added to monolayers of mouse embryonic cells, a characteristic cytopathic effect and cell death are observed. If the virus is diluted, viral infection of suitable cultures causes lysis of cells leaving holes in the monolayers. These *plaques*, as they have been called, have proved useful in assaying viral suspensions quantitatively. Monolayers of mouse cells are permissive for polyoma; they allow infection and production of viral particles and the cells die by lysis. Whereas cytopathic effects are observed in tissue cultures of mouse embryo cells, hamster cells are transformed by polyoma virus. The transformed cells produce tumors when inoculated *in vivo*. Thus, while certain cells support viral replication (permissive for the particular virus) and usually die as a result, others are transformed and survive.

C-viruses that are found in apparently normal cells sometimes cause cancer when injected into a suitable host (Todaro and Heubner, 1972). Thus, viruses may cause cell death, transform cells, or exist as an apparently innocuous passenger in a cell.

Stoker and MacPherson (1961), in devising plating techniques for clonal analysis of transformation, added dilutions of polyoma virus to suspensions of baby hamster kidney cells (BHK-21). These cells were heteroploid and had an increased growth rate and cloning efficiency over the normal, but they were believed to be similar to normal cells in other respects. The technique allowed for quantitation that had previously been impossible. However, Defendi, Lehman, and Kraemer (1963) showed that BHK-21 cells of apparently normal appearance could, upon transplantation into suitable hosts, produce malignant neoplasms with high frequency. Thus, there appears to be a poor correlation between characteristics of viral transformed cells and tumorigenicity; some cells, as in the case of BHK-21, may lack features of transformation and be tumorigenic and others may appear transformed and lack tumorigenicity.

Virally transformed cells have antigens not found in the normal cells. Those that are present in the cell membrane (tumor specific transplantation antigens) can elicit a graft rejection response by the host. For example, a plate of fibroblasts from a C_3H mouse, transformed with SV_{40} virus, will produce tumors in C_3H mice only if the host's immune system is suppressed by treatment with x-ray or cortisone.

In addition to the transplantation antigens, there are intracellular antigens that are encoded in the viral genome and can be used as markers of viral infectivity. These are the tumor antigens (T antigens) of DNA viruses and the gs antigens of RNA viruses.

DNA tumor viruses

Oncogenic DNA tumor viruses can be classified into three major groups: papovaviruses, adenoviruses, and herpesviruses. The papovaviruses include polyoma, SV_{40}, and papilloma virus. Polyoma virus and SV_{40} virus have been extensively studied because their genomes are small and they encode for only a few proteins. Thus, it should be easier to identify essential transforming genes in them than in larger viruses. The polyoma virion (extracellular nature virus) is only 45 millimicrons in diameter, spherical, and surrounded by protein capsomeres arranged in icosahedral symmetry. The virus contains 12%, by mass, circular double-stranded DNA and 80% protein; the latter are mostly polypeptides of the capsid coat. There are from 2.9 to 3.5 $\times$ 10^6 daltons of DNA, enough to code for 4 proteins of 50,000 molecular weight or 10 proteins of 19,000 molecular weight. The relationship of many of the genes on the DNA has been mapped. Similar information is available for SV_{40} virus (reviewed by Tooze, 1973).

It is a general rule that infection with these oncogenic viruses is either productive (permissive for virus replication) or transforming. A single viral particle can transform a cell. The polyoma virus is tumorigenic in hamsters, mice, and rats and is permissive in mouse embryo cells, while SV_{40} is tumorigenic in hamster, mouse, and human cells and is permissive in monkey kidney cells. Less is known about the papilloma viruses that are tumorigenic in man, rabbits, and dogs. According to Green (1966), the following occurs during productive infection with polyoma virus, *e.g.*, in mouse fibroblasts:

1. Transport of partly uncoated virus to the nucleus.
2. Uncoating of the virus.
3. Transcription of parts of viral genome to form early messenger RNA (mRNA).
4. Translation of early mRNA to early viral proteins.
5. Replication of viral DNA.
6. Transcription of late mRNA.
7. Translation of late mRNA to viral structural proteins.
8. Assembly of virus from viral DNA and viral structural proteins.
9. Lysis of cell with release of virus.

In transforming infections, the virion crosses the cell membrane in pinocytotic vacuoles and is uncoated (manner unknown) in the cytoplasm. The double-stranded DNA containing the viral genome is then integrated into the DNA of the host if transformation is to take place.

Proof that the viral genome is integrated into that of the cell is derived from several experimental approaches. In the first approach, *rescue*, the transformed cells (which do not produce virus) are hybridized to permissive cells; the resultant hybrids then produce the DNA virus. In the second approach, purified nuclear DNA from transformed cells, when incorporated by permissive cells, results in the production of virus. Finally, since mRNA is a direct transcriptional copy of its DNA, it has been possible to anneal DNA from transformed cells to labeled mRNA of viral origin, thereby demonstrating the complementarity of their base sequences. These studies indicate the presence of the viral genome in the DNA of the cell.

After integration, viral directed messenger RNA is produced, enters the cytoplasm, and results in the synthesis of tumor antigen (T antigen) that is quickly transported into the nucleus. T antigen is an antigen of approximately 70,000 molecular weight, but its function is not known. It is associated with chromatin. In transforming situations the late functions of the viral genome are not expressed, which may account for the lack of production of virus. The factors involved in blocking viral production and causing transformation are not known.

Because the late gene functions leading to production of the virion proteins are blocked, a search has been made in the early events after infection for the factors controlling transformation. Temperature-sensitive mutants * of the early genes have been produced experimentally. Tegtmeyer (1972) studied temperature-sensitive mutants of SV_{40} and discovered that there is a gene in SV_{40} necessary for initiation of viral DNA synthesis and transformation. Whether or not it is a specific transformation gene has not been determined.

Infection with polyoma induces synthesis of cellular DNA in confluent primary cultures (Dulbecco, Hartwell, and Vogt, 1965; Winocour, Kaye, and Stollar, 1965), but the mechanism by which normal cellular controls are altered is unknown. It is probable that stimulation of DNA synthesis plays an important role in transformation. Since there is insufficient genetic information in the virion to account for the number of enzymes required, the virus must induce the host cell to produce them.

Viruses with a morphology similar to that of the polyoma virus and SV_{40} have been isolated from brain tissue of patients with massive demyelinating diseases, called *progressive multifocal leukoencephalopathy*, and from the urine of immunosuppressed patients. Some of these are oncogenic in lower species, but it is not known if they produce tumors in man or even if they produce the demyelinating diseases (Padgett *et al.*, 1971; Weiner *et al.*, 1972).

Most of the adenoviruses cause respiratory diseases in their natural hosts, but when the virus is used to infect another species, it may be tumorigenic. For example, human adenovirus 12, which causes a pneumonitis in humans, can cause tumors in newborn rats and other rodents. The adenoviruses are more complex than the papovaviruses (Ginsberg *et al.*, 1966). The particle is approximately 80 mμ in diameter and is covered by a protein coat with icosahedral symmetry. Extending from specific bases in the capsid are fiber proteins capped by knobs. The DNA is from 10 to 25 $\times$ 10^6 daltons and could code for from 25 to 50 proteins of approximately 30,000 molecular weight. In a lytic cycle of infection the virus is adsorbed, penetrates, and is partially uncoated. The DNA enters the nucleus and transcription begins. A cellular RNA polymerase transcribes an early viral mRNA. This RNA is translated in the cytoplasm and a T antigen and several other proteins are synthesized. The synthesis of viral DNA occurs in the nucleus, and replicative intermediates with single stranded DNA have been

---

* Temperature-sensitive mutants exhibit wild type behavior at permissive temperature and abnormal behavior at nonpermissive temperature.

described. Later, as viral DNA synthesis proceeds, there is transcription of a late set of viral genes. Most of the host cell's synthesis of DNA, RNA, and proteins is markedly reduced while there is a marked increase in the synthesis of viral proteins. The virion proteins are transported to the nucleus where the virus is assembled.

In cells that are transformed by adenoviruses, molecular hybridization and studies on the rate of renaturation of adenovirus DNA in the presence of nuclear DNA have been used to detect integrated sequences of adenovirus DNA in the nuclear DNA. Green *et al.* (1970) have demonstrated that in adenovirus transformed cells, only a fraction of the adenovirus DNA is being transcribed. It is possible that not all of the genes of the adenovirus are integrated into the host DNA. Extensive studies are in progress to map the adenovirus genome and to understand the function of certain adenovirus genes by using temperature-sensitive mutants.

The herpesviruses are the cause of some animal cancers. A herpesvirus that causes lymphomas when inoculated into monkeys has been isolated from squirrel monkey kidney cells and a herpesvirus that causes lymphocytic leukemia in guinea pigs has been isolated from guinea pig cells. In chickens there is a disease called *Marek's disease* that appears to be caused by a herpesvirus (Lee *et al.*, 1971). It is characterized by infiltrations of nerve trunks and other organs by lymphocytes. A herpesvirus seems to be the cause of Lucké renal adenocarcinoma in the frog. Herpes simplex-2 virus may play an etiological role in human cervical carcinomas. Similarly, the EB (Epstein–Barr) virus may be the cause of Burkitt's lymphoma, nasopharyngeal carcinoma, and infectious mononucleosis in man (Zur Hausen and Schulte-Holthausen, 1970).

The herpesviruses have a lipid glycoprotein envelope that contains several glycoproteins. The DNA is approximately 100×10^6 daltons, linear, and double-stranded. During infection the virus is adsorbed to cellular receptors and then penetrates the cell either by fusion with the cellular membrane or by pinocytosis. The virus is uncoated and the DNA-protein complex enters the nucleus where it is transcribed. The pattern of specific viral mRNAs changes during the infection. Viral proteins are synthesized in the cytoplasm and migrate to the nucleus where they assemble with the replicated DNA. The virus is then enveloped by a membranous coat at the inner surface of the nuclear membrane. It is not clear exactly how the virus leaves the cell. The cells that allow permissive replication of the virus die, but in transformed cells the viral DNA (or at least a part of it) is integrated into the host's genome. This DNA contains information for certain viral antigens detectable in the herpesvirus-transformed cells.

RNA tumor viruses

The RNA tumor (or oncorna-) viruses include leukosis viruses that cause a leukemia-like disease and sarcoma viruses that are capable of transforming fibroblasts in culture or producing sarcomas in animals. They can cause malignancies in fowl, rodents, and cats. The avian leukosis and sarcoma viruses have been the most extensively studied. The virion contains a 70S RNA that is approximately 10×10^6 daltons. It consists of 35S subunits thought to be the basic genetic component (Erikson, 1969; Wang *et al.*, 1976) and several species of low molecular weight RNA, predominantly 4S transfer RNAs (Erikson and Erikson, 1976). The low molecular RNAs are derived from the cell.

For many years investigators searched for the mode of replication of these viruses and the means by which their genetic information was passed to their progeny. Temin and Mizutani (1970) and Baltimore (1970) found that the virion of oncornaviruses contained an RNA-dependent DNA polymerase that converted the genetic information in RNA into DNA (reverse transcriptase). Subsequently, it was found that the DNA synthesized from the RNA template was integrated into the genome of the infected cell. In contrast to the situation with DNA tumor viruses, cells infected and transformed with RNA tumor virus frequently continue to produce viruses.

Studies are in progress to characterize and isolate the protein products from the RNA tumor virus. Several of the virion proteins, including the structural and the reverse transcriptases, appear to be coded for by the viral RNA. Recent studies by Bishop and Varmus (1975) suggest that there is a portion of the viral genome that contains the information for sarcomatous transformation of cells and studies are in progress to determine what controls the transcription of this portion of the genome. Temperature-sensitive mutants and deletion mutants of Rous sarcoma virus have been isolated and used to map the viral genes. Cells transformed by certain temperature-sensitive RNA tumor viruses can be modulated to express a normal or transformed phenotype by changing the temperature of the cultures. The use of defective virus is difficult to study in systems where helper viruses are required for replication of the sarcoma virus. (Helper viruses are oncornaviruses that are not capable of transforming fibroblasts in culture and are frequently found in normal cells (Hanafusa and Hanafusa, 1966).

There are interesting relationships between exposure to chemical carcinogens and oncornaviruses. In rat cells in culture, the efficiency of transformation upon exposure to both a chemical carcinogen and a sarcoma virus far exceeds the transformation observed when either agent is used alone. Normal cells contain the genome of oncornaviruses, but it is not known if this information is used by the cell. Possible exogenous agents such as carcinogens may modify or activate the expression of this genetic information and result in cancer. This is in line with the concept that the genetic information for neoplasia resides within the host's own genome, and its expression involves the mechanism of differentiation.

The environment and cancer

Man is a species capable of making major changes in the environment. Such changes have been most evident since the Industrial Revolution and have been markedly accelerated during the last decade. The unsought and even harmful byproducts of technical advances have, until recently, been accepted as the price of progress. Now, concern about the deleterious effects of environmental alterations has become a major social issue. Much of this concern is directed at the carcinogenic hazards (Berg, 1976; Fraumeni, 1975; Hammond, 1975; Schottenfeld, 1975), real or potential, of such factors as food additives, pesticides, polluted air and water, medications, and radiation.

The concept that environmental factors play a role in human carcinogenesis is not new (Shimkin, 1975). In the late eighteenth century Sir Percival Pott pointed out the relationship between scrotal cancer, soot, and the occupation of chimney sweeping. However, the demonstration during the nineteenth century of relationship between exposure to arsenic, coal tar, and skin cancer and of relationships between aromatic amines and bladder cancer in dye workers has not yet resulted in the elimination of these hazards. Nor is radiation-induced cancer entirely a product of the nuclear age. A lethal lung disease, presumably cancer, was prevalent among miners in Central Europe in the sixteenth century, and in the nineteenth century a high incidence of lung cancer was found in Schneeberg miners. As in today's Colorado uranium miners, the high incidence of lung cancer is at least partially ascribed to the inhalation of radioactive dust. The high incidence of skin cancer in persons exposed to excessive sunlight (sailors, farmers) has been known for 200 years, and the relationship between x-rays and skin cancer, and later, leukemia, became known not long after Roentgen's discovery of x-rays.

With few exceptions, clear-cut associations, such as those cited, between carcinogen and human cancer are, now, far more difficult to document because of the ever increasing number of variables in an environment of increasing complexity. It has been estimated that nearly 2 million organic chemicals have been synthesized, and 250,000 are added each year. Of this great number, less than 1% have been tested for carcinogenicity, and testing may have been inadequate for many of these (Carter, 1974). Many chemicals are virtually ubiquitous and are used as food additives (Kermode, 1972). The food coloring agent, Red Dye #2, has been used for almost a century. Originally synthesized from coal tar, it has more recently been derived as a petroleum by-product. After many years of experimentation had produced conflicting results as to the carcinogenicity of Red #2 in animals, and after no evidence that Red #2 caused human cancer, the dye has been banned from further use. Its banning seems to have resulted more from frustration than documentation. Even less is known about its potential replacement, Red #40.

Warnings regarding the potential carcinogenicity of chemicals, based on experimental data, have gone unheeded and have been subsequently shown to be valid. Vinyl chloride, the precursor for the ubiquitous food wrapping, polyvinyl chloride, was shown to be carcinogenic in rats in 1971. Several years later, a rare form of cancer, hemangiosarcoma of the liver, was described in workers exposed to vinyl chloride (Wegman *et al.*, 1976). Chloromethyl methyl ether, widely used industrially as an intermediate in organic synthesis, was shown to be carcinogenic in mice and rats. A marked increase in lung cancer has been found in workers exposed to chloromethyl methyl ether (Figueroa, Raszkowski, and Weiss, 1973).

Does the fact that an agent causes cancer in experimental animals mean that it will do the same in man? The answer is yes, no, and maybe. Phenobarbitone induces hepatic microsomal enzymes (for detoxification of poisons) and causes liver cancer in mice. Phenobarbitone has been used for many years in the long-term therapy of seizures, and individuals treated for many years with this drug have a lower incidence of cancer in general instead of an increased incidence of liver tumors. What if a particular agent fails to cause cancer in the particular species, dose, and time allotted to the testing situation? Does this mean that it will not cause cancer in humans? Rehn (1895) first reported the occurrence of bladder cancer among workers in the German analine dye industry, an incidence that has been reported as up to 60 times that of unexposed persons. Repeated attempts to induce bladder cancer, using various dye components, *e.g.*, β-naphthylamine, in experimental animals were unsuccessful. As has been previously discussed, it was eventually determined that the ultimate carcinogen was not β-naphthylamine but its metabolic derivative and that whereas humans had the biochemical machinery necessary for this metabolic conversion, mice did not. Finally, it was shown that where the human metabolite, 2-amino-1-naphthylamine, was incorporated into wax pellets and inserted into the bladder of mice, it did cause bladder cancer. Thus, species differences make the application of animal data to the human situation unreliable (Kuschner and Laskin, 1971).

A variety of screening procedures have been suggested to identify chemicals that may be carcinogenic, but none is ideal (Saffiotti, 1971; Bridges, 1976). Laboratory rodents are not the whole solution because most chemicals must be metabolized to active carcinogenic forms, and there is no assurance that the test animal has the appropriate enzymes. The same limitation applies for testing in tissue culture. Ames *et al.* (1973) has proposed a screening procedure to detect chemicals mutagenic in bacteria. Although there is a correlation between mutagenicity and carcinogenicity, the relationship may be fortuitous and fraught with significant exceptions. At this time the most reliable test system for determining the carcinogenicity of chemicals in human beings is man.

A complicating factor in the study of carcinogenesis in man is the long latent period between exposure to an agent and the appearance of clinical cancer. The latent period for bladder cancer among workers at risk is approximately 40 years (Hoover and Cole, 1973). Cigarette smoking (Hammond, 1975; Wynder and Hoffman, 1976) was not known to be a cause of lung cancer when this habit became so popular; it was not until 30 and 40 years later that the relationship became clear.

If carcinogens of the future are with us now, with our remarkable technology and scientific resources, why can't we identify and eliminate them? The enormity of the problem is illustrated by the class of compounds known as the *nitrosamines* (Shapley, 1976). Nitrosamines are well documented carcinogens in a wide variety of animals and result in a wide spectrum of cancers; in low doses they may act as co-carcinogens. They are virtually ubiquitous, particularly in urban areas, and are found as decomposition products in air, soil, water, and sewage. Their precursors include nitrogen and nitrous oxides (the principal products of most fuel combustion), amines (found in pesticides, drugs, and foods), and nitrates and food additives (used to color preserved meat and fish). How does one determine from such a bewildering array of data the potential carcinogenic effects in humans of the nitrosamines and their precursors? And, if such effects were found, how could they be eliminated—short of completely altering almost every aspect of our environment? Even if the carcinogenic effects of such agents are identified, society will have to determine if the risk is a worthwhile price to pay in return for the beneficial effects. Or, as in the case of automobile emission controls, attempts may be made to reduce the carcinogenic effect without eliminating it.

On the other hand, some risks might be identified and eliminated more easily. Hepatocellular cancer is rare in the United States and Europe, but it accounts for as much as 65% of all cancer in some parts of Africa and Asia. In the early 1960's an outbreak of this cancer occurred in trout in the northwestern United States. This was traced to contamination of their food by aflatoxin, a product of the fungus, *Aspergillus flavus*, an observation that was confirmed experimentally. Elimination of the carcinogen resulted in elimination of the tumor. Aflatoxin (and *Aspergillus*) are found in mouldy maize and peanuts stored out-of-doors in warm humid climates characteristic of the locales where hepatoma is common (Swenson, Miller, and Miller, 1973). Surveys in Thailand have shown that the incidence of hepatocellular carcinoma is related to the quantity of aflatoxin consumed. If aflatoxin is the source of this most common form of cancer in these parts of the world, its elimination would appear to be relatively easy, at least when compared to nitrosamines in urban societies.

It has been suggested that dietary factors play a role in certain forms of cancer (Higginson, Terracini, and Agthe, 1975). Carcinoma of the stomach is common and carcinoma of the colon is uncommon in Japan—the reverse of the situation that exists in the United States. Studies of Japanese immigrants to Hawaii (and their progeny) show a gradual change in the incidences of these two forms of cancer to one approaching that of other Americans. The change in diet—to one approximating the American diet—has been correlated with this alteration in cancer incidence (Wynder and Reddy, 1973).

There are a number of other situations in which the population is sufficiently unique, or the agent specifically defined, or the resultant cancer ordinarily so rare, such that the cause-and-effect relationship is apparent. Human cancers associated with radiation illustrate some of these points (Miller, 1975; Jablon, 1975). There is a high incidence of bone cancer in women who paint watch dials with radium, pointing the brush with their lips and ingesting minute amounts of radium. Therapeutic radiation has been followed by leukemia in persons treated for a form of rheumatoid arthritis affecting the spine and in thyroid cancer among adolescents and young adults treated during infancy for what was then thought to be a pathologic enlargement of the thymus gland. The survivors of the Hiroshima and Nagasaki atomic blasts have had increased incidences of leukemia (Anderson, 1971), lymphoma, thyroid, and other forms of cancer, and it has been suggested that the offspring of mothers who underwent diagnostic x-ray procedures during pregnancy have a greater than expected incidence of cancer in childhood.

A number of drugs have been associated with carcinogenesis. Metronidazole was until recently widely used as the most effective treatment of trichomoniasis, a common protozoal infection of the vagina. Based on its demonstrated carcinogenicity in animals it was withdrawn from the market. Estrogens, as we have seen, are tumor promoters in tissues sensitive to their hormonal effects, an observation that has been amply demonstrated experimentally and corroborated by certain clinical situations. For example, there is a high incidence of endometrial hyperplasia or carcinoma in women with estrogen-secreting ovarian tumors. The widespread use of such agents in oral contraceptives, as well as in the treatment of infertility and of postmenopausal symptomatology, has led to concern regarding their potential carcinogenic effects in the millions of women so treated (Marx, 1976). Although no association between the use of such agents and cancer had been observed, their relatively recent application and the possibility of a long latent period before the appearance of such tumors have precluded assurance of their safety.

Contamination of the air with various chemicals has been of major concern, particularly with increased urbanization, industrialization, and use of the automobile. The polycyclic aromatic hydrocarbons, whose efficacy in producing cancers in animals is well known, have been of particular concern, with benz(a)pyrene the agent most suspected. The occupational hazards associated with long-term exposure to high concentrations of such carcinogens as arsenic, asbestos, and vinyl chloride have been well documented, but what of nonworkers in the surrounding community who are exposed, often without their knowledge, to lower doses of the same carcinogens? Blot and Fraumeni (1975) found a significantly increased mortality due to lung cancer in locales in which smelting and refining of ores resulted in the release of airborne inorganic arsenic.

Although specific environmental factors may not occur in quantities sufficient to result in cancer, occurring together they may act as co-carcinogens. Selikoff, Hammond, and Churg (1971) examined a large group of asbestos insulation workers, noting that previous studies had shown such individuals to be 7 times more likely to develop lung cancer than matched peers. Further analysis of this population showed 2 subpopulations: nonsmokers, none of whom died of lung cancer, and smokers, who had a risk of dying of lung cancer 8 times greater than that of smokers not exposed to asbestos. Those who were both exposed to asbestos and smoked had a likelihood of developing lung cancer nearly 100 times greater than similar men who neither smoked nor were exposed to asbestos.

Finally, in considering environmental factors and cancer, the possible role of infectious agents must be mentioned. In chickens, a lymphoma-like disease, Marek's disease, has been shown to be due to a herpesvirus and can be prevented, like some other non-neoplastic infectious diseases, by vaccination (Purchase, 1976). Feline leukemia and lymphoma, the most common forms of cancer in domestic cats, is due to a "C-type" or oncornavirus that is horizontally transmitted via secretions, *i.e.*, it is an infectious disease in this species. Although the virus can be grown in human tissue culture cells, and there has been concern about its potential oncogenicity in humans, neither the virus nor its antigens have been identified in human tumors (Levy, 1974). Nevertheless, the possibility of animals, domestic or wild, as reservoirs or vectors of some types of human cancer exists.

The evidence for a viral etiology, let alone a transmissible viral etiology, for human cancer has not been established. Epidemiologic evidence in Burkitt's lymphoma (*e.g.*, the fact that its geographic incidence is related to altitude, humidity, and temperature) suggested an infectious etiology and a virus, the Epstein-Barr virus, has been isolated from these tumors. The same virus has been found in nasopharyngeal cancers and in patients with infectious mononucleosis, and, recently, it has induced tumors in marmosets (Henle and Henle, 1974; Klein, 1975).

"Outbreaks" or time-space clusters of Hodgkin's disease (Albany, N.Y.) and leukemia (Niles, Ill.) have been reported and used as evidence for an infectious etiology for these diseases; whether such outbreaks are statistically significant has been questioned (Smith and Pike, 1976). Evidence has been presented to support the concept that carcinoma of the cervix is an infectious (*i.e.*, venereal) disease (Kessler, 1976). For example, its statistical association with early age of first intercourse, multiplicity of sex partners, etc., and seroepidemiologic studies have implicated Herpes type-2 as an etiologic agent (Melnick, Adam, and Rawls, 1974). A major portion of cancer research today is devoted to the search for viruses in human cancers with the hope that the identification of such agents would, as in the case of Marek's disease in chickens, be followed by the development of vaccines.

In summary, despite advances in early diagnosis and therapy, the cost of cancer remains unacceptable. Screening of agents either by the use of experimental animals or *in vitro* testing may or may not have relevance to their ability to cause human cancer. We must identify, by careful prospective monitoring and/or retrospective analyses, agents and situations that are obviously carcinogenic in humans. However, it is becoming increasingly apparent that in an environment that is increasingly complex and artificial, the elimination of all such agents from the environment may be socially and economically unacceptable.

SEVEN

Genetics and Neoplasia

Inheritance

Animals

There is a wide range in the incidence and types of cancer among those few species, subspecies, and breeds for which data are available (Meier, 1963). The situation in dogs and cats is an example. Although cancer is nearly twice as frequent in dogs, the proportion of neoplasms that are malignant is much higher in cats (Brodey, 1970). Mast cell and testicular tumors are common in dogs and rare in cats, while lymphosarcomas occur five times more frequently in cats. There is marked variation among breeds of dogs (Mulligan, 1949). The boxer has a predilection for many tumors; these tumors are of a different type from those that occur in the cocker spaniel, and the beagle has a low incidence of tumors in general.

The development of inbred strains of mice has resulted in a clear-cut demonstration of genetic variation in incidence of neoplasms (Heston and Vlahakis, 1967). More than 95% of breeding female C_3H strain mice develop mammary carcinomas, while only 1% of similar C57 strain mice develop this tumor. Leukemia, which is rare in C_3H and C57 mice, occurs in more than 50% of AKR mice. Lung tumors, common in BALB/c mice are rare in C57, C_3H, and AKR strains.

Such marked intraspecies variation in the incidence of neoplasms among genetically distinct populations was accepted by many investigators as a demonstration of genetic predominance in the determination of cancer. However, Bittner's observations of concentration of mouse mammary cancer viruses in high mammary tumor strains indicated the importance of extragenetic factors (Bittner, 1939).

The development of mammary cancer in mice appears to be multifactorial. There is an event determined by a virus influenced by suitable genetic and endocrine factors. The role of endocrine factors in their genesis is evident since the tumors occur most frequently in females that have been used for breeding. Genetic factors might act at a variety of levels, *e.g.*, providing a favorable environment for the maintenance of the virus and its transmission to offspring, providing a favorable situation for the cellular transformation of mammary tissue to the neoplastic state, or for the progression of an established neoplasm. Such considerations are speculative because these genetic factors have not been characterized. High incidence strains presumably have resulted from a fortuitous concentration of genes favorable to the development of certain tumors, but they are nevertheless, still subject to the effects of nongenetic factors.

Mammary tumors in inbred strains of mice illustrate the concept of inheritance of threshold or quasicontinuous characteristics, *i.e.*, phenotypic traits whose appearance represents the culmination of many genetic and nongenetic determinants (Heston, 1963). Selective breeding for such a trait (*e.g.*, mammary cancer) should maximize the incidence determined by genetic factors with a range of variation determined by nongenetic factors. Conversely, one might attempt to select for and develop strains with fewer genetic factors and, ultimately, a single genetic factor that influences the likelihood of cancer. Some host genetic factors might confer resistance to cancer development and result in low incidence strains. Axelrad (1969) has identified eight genetic loci that influence the response of mice to a leukemogenic virus; one of these loci has an allele that confers host resistance to the virus.

The role of multiple genetic factors in the development of cancer is illustrated by melanomas in the hybrid offspring of platyfish and swordtails (Anders, 1967). Platyfish have genetically determined patterns of pigmented spots. These patterns are determined by a major *color* gene and a variety of modifier genes. The pigment pattern of the swordtail is also genetically determined. One pattern of platyfish pigmentation is the presence of black spots on the dorsal fin; this results from the presence of a specific Sd-color gene and certain modifier genes. When Sd-bearing platyfish are mated with swordtails, many of the F-1 hybrids develop accelerated and excessive growth of pigment spots that become slow-growing melanomas. Backcrossing of these F-1 hybrids with swordtails increases the likelihood of melanomas that grow rapidly and spread widely. Anders suggests that these breeding experiments result in the dilution of platyfish Sd-modifier genes that normally control the growth of pigment spot cells and that the resultant cancer is caused by a genetic imbalance. It is also conceivable that oncogenic viruses may be activated in these experiments.

Human beings

A high frequency of cancer has been reported over multiple generations in a number of so-called *cancer families* (Lynch *et al.*, 1976). In these families a variety of types of cancer are found (although tumors of one or several organs may predominate). For example, family "G," first reported by Warthin in 1913, and more recently by Lynch and Krush (1971), has shown a high frequency of cancer. Family members tend to develop cancer at an early age and often have multiple tumors. Marked variation in the frequency of cancer among family branches was felt to be consistent with an autosomal dominant pattern of inheritance, but again, variable penetrance * suggests a quasicontinuous condition determined by both genetic and nongenetic factors.

Very few human tumors occur with a familial frequency sufficient to suggest a simple Mendelian pattern of inheritance (Knudson, Strong, and Anderson, 1973). One of these is retinoblastoma, a rare malignant tumor of the eye that occurs in children. Although most are sporadic, about one-third are familial and appear to be inherited as an autosomal dominant with variable penetrance. Recently, retinoblastoma was found in 7 of 12 patients with deletion of the short arm of a D chromosome, suggesting that the genetic locus is located at that site (Taylor, 1970). The association of a cancer with a specific chromosomal abnormality is the exception rather than the rule, and none of the other inherited forms of cancer has been associated with demonstrable specific chromosomal abnormalities. The variable penetrance of retinoblastoma again suggests a quasicontinuous genetic character whose expression depends upon other factors. These factors are unknown.

There are rare examples of other tumors that are inherited as isolated phenomena or as a part of a syndrome that includes non-neoplastic features (Knudson, Strong, and Anderson, 1973; Anderson, 1970; Lynch *et al.*, 1976). In familial polyposis, which is inherited as an autosomal dominant, the large intestine contains hundreds of benign polyps. Invariably, adenocarcinoma of the colon develops in these individuals. Nevoid basal cell carcinoma syndrome is inherited as an autosomal

* *Variable penetrance* is the deviation of the incidence of an inherited trait within a pedigree from that expected on the basis of Mendelian genetics.

dominant with variable penetrance and expressivity. Among the various components of this syndrome is the occurrence of numerous, sometimes hundreds, of basal cell carcinomas of the skin (Jackson and Gardere, 1971). In neurofibromatosis, another autosomal dominant, multiple tumors are the major finding.

Of particular interest are rare, inherited syndromes characterized by both chromosomal instability and a tendency to develop cancer. These have been termed the *chromosomal breakage syndromes* (German, 1972). Bloom's syndrome is inherited as an autosomal recessive and occurs predominantly in Jews of eastern European extraction. It is manifest early in life by short stature, a typical and peculiar facies, and a facial rash. Among 45 cases studied by German, 9 developed some form of cancer, usually at a young age. Somatic cells derived from these patients have an increased frequency of nonspecific chromosomal breakage and rearrangement. An association between chromosomal fragility and a propensity to develop cancer has been described in 2 other inherited syndromes, Fanconi's anemia and ataxia telangiectasia. The significance of the relationship of inheritance, chromosomal instability, and cancer in these inherited syndromes has not been elucidated, but it is of interest that these cells are more easily transformed with SV_{40} than are normal cells.

Genetically determined biochemical defects in DNA metabolism have been described in xeroderma pigmentosum, another inherited syndrome associated with increased cancer risk. This rare autosomal recessive disease is characterized by onset of abnormal skin pigmentation early in life followed by various malignancies in parts of the skin that are exposed to sunlight. Fibroblasts from these patients maintained *in vitro* repair ultraviolet-induced DNA damage slowly or not at all. Biochemical and cytogenetic analyses suggest that xeroderma pigmentosum may actually represent a spectrum of disorders, each reflecting a genetically determined defect in different DNA-repair enzymes (Robbins *et al.*, 1974). Study of xeroderma pigmentosum is particularly promising as a means of elucidating relationships between genomic and environmental factors in the genesis of neoplasms.

For many years clinicians, epidemiologists, and geneticists have searched for genetic patterns in the more common forms of human cancer. Carcinoma of the breast occurs in approximately 5% of American women. One would expect any genetic pattern for so common a tumor to be readily apparent, but familial tendencies that are not Mendelian are difficult to interpret. In reviewing the family histories of more than 6,000 cases of breast cancer, Anderson (1972) found 500 patients who had at least 1 close relative with breast cancer. Those patients tended to form 2 subtypes with regard to a variety of parameters, such as age of onset, bilateral breast involvement, and blood group. The incidence of bilateral breast cancer was three times higher in patients with a positive family history. In the subtype composed of those in whom the cancer occurred before menopause, bilaterality was five times more frequent than in the negative family history group. These data suggest a genetically determined influence on the behavior of the tumor.

Reports of familial tendencies to develop specific cancers are numerous. Among 154 patients with endometrial carcinoma, Lynch *et al.* (1966) found 20 with mothers or sisters who had the same disease. Tokuhata (1964) reported that an individual who smokes cigarettes and has a close relative with lung cancer has a fourteenfold increased risk of developing lung cancer—a risk much greater than that associated with cigarette smoking alone. Rigby *et al.* (1968) found that of 151 patients with leukemia, 39 had one or more family members with leukemia or lymphoma. Innumerable reports cite instances of the same type of tumor in two or more siblings or in twins.

The incidence of cancer, and of certain specific forms of cancer, varies greatly among different populations. Carcinoma of the breast, for example, is uncommon in Orientals, common in Americans, and very common in Parsi women in India. The dangers of ascribing a high incidence of a particular form of cancer to genetic factors alone are illustrated by investigations of breast cancer in the Parsi (Paymaster and Gangadharan, 1972). The Parsi are a religious sect of approximately 80,000 persons comprising about 2% of the population of Bombay. They are forbidden by their religion to marry outside the sect and have maintained this genetic isolation for 13 centuries. Although cancer is a relatively infrequent disease among these people, cancer of the breast accounts for approximately 50% of all cancers in Parsi women. Mammary carcinoma occurs 2 to 3 times more frequently among the Parsi than among other residents of Bombay. Is this form of cancer inherited among these genetically isolated people, or is this an example of vertical transmission of an etiologic agent in genetically predisposed hosts (*e.g.*, as in inbred mice)? Recently, viral particles, similar to those of mouse mammary tumor virus, have been discovered in the milk of a high percentage of Parsi women (Moore *et al.*, 1971).

Chromosomal abnormalities in cancer patients

We have already alluded to the high frequency and early age of the onset of cancer in individuals with the inherited *chromosomal breakage* syndromes (Bloom's syndrome; Fanconi's anemia) and the association between retinoblastoma and partial deletion of a D group chromosome. Patients with trisomy 21 (Down's syndrome) have an increased incidence (from fourfold to twentyfold) of leukemia. A rare tumor, gonadoblastoma, is most often seen in the testis of undeveloped gonads of individuals with certain sex chromosome abnormalities.

However, Harnden (1970), in a survey of a large number of unselected cancer patients, found no evidence of an increased incidence of karyotypic abnormalities. Bottomley, Trainer, and Condit (1971) found no karotypic abnormalities in a "cancer" family. Such negative findings require reexamination using banding pattern analysis.

A variety of chromosomal abnormalities is seen in many human cancers, experimental and spontaneous animal cancers, and transformed cells. Alterations in chromosome number and morphology, and their possible significance, have been reviewed many times in recent years (DeGrouchy and DeNava, 1968; Lampert, 1971; Atkin, 1974; German, 1974; Nowell, 1974; Sandberg and Sakurai, 1974; Makino, 1975; Hirschhorn, 1976). One generalization that may be made about chromosomal abnormalities in cancer is that they are highly variable. Even tumors of the same histologic type, arising in the same organ within animals of the same species, show such a range of karyotypic variation that no two tumors are karyotypically identical. Even within an individual tumor there is usually some degree of karyotypic variation and this may increase or decrease during the life of the tumor. This karyotypic instability is typical of human malignancies.

Number

Many animal tumors and the vast majority of human cancers are aneuploid, *i.e.*, they are characterized by a mode different from the normal diploid number characteristic of the species. Of the few benign tumors examined, some have been shown to be aneuploid as well; on the other hand, some human malignancies (*e.g.*, 50% of acute leukemias) are diploid. Nevertheless, nearly all other human malignancies are aneuploid. In one survey of 21 solid human cancers (Yamada, Takagi, and Sandberg, 1966), all were aneuploid; in fact, the investigators were unable to find a single diploid cell in any of the tumors. Aneuploid tumors show no characteristic modal number that can be said to be representative of a particular histologic type or tissue of origin. Although some tumor types show a tendency to hover about a particular range of modality, examining a tumor's chromosomes is useless when attempting to distinguish one type of tumor from another. A wide range from the mode is sometimes correlated with a more anaplastic appearance and malignant behavior, but this association is not reliable.

Morphology

The chromosomes of cancer cells may vary as much in size and shape as they do in number, and most cancer cells contain at least one such deviant chromosome. Morphologically abnormal chromosomes have been identified in some diploid cancers—giving rise to the term *pseudodiploid*. In some tumors, many or most of the cells contain the same aberrant chromosome. Such *marker* chromosomes may distinguish an individual tumor, but not a type of tumor. Thus, each of five colonic carcinomas might show a marker chromosome, but the marker is different in each tumor.

The only exception to this nonspecificity of marker chromosomes is seen in chronic myelogenous leukemia. In more than 90% of the cases virtually all of the myelogenous stem cells contain the distinctive Philadelphia chromosome. This is chromosome #22, a chromosome originally thought to result from a partial long arm deletion, but now known to be due to an unbalanced translocation (Rowley, 1973). The Philadelphia chromosome is seen in the stem cell precursors of erythrocytes and platelets as well as granulocytes, but not in any other cells from patients with chronic myelogenous leukemia. Occasionally it has been demonstrated before the clinically overt appearance of leukemia (Nowell, 1971). In a few other neoplasms, distinctive chromosomes have been found in a small proportion of tumors of the same type; in none, however, does the reliability approach that of the Philadelphia chromosome in chronic myelogenous leukemia.

Analysis of chromosome morphology, of normal as well as neoplastic cells, has been limited by available techniques, and until recently only major alterations in chromosome structure have been demonstrable. The introduction of chromosomal banding by Caspersson, Lomakka, and Zech (1971) has permitted the identification of individual chromosomes and the detection of relatively minor additions, deletions, and rearrangements. This major advance in cytogenetics is now being applied to the investigation of neoplasms (Wurster-Hill, 1975).

EIGHT

Controls

Control of malignant expression

Approaching carcinoma clinically, developmentally, biochemically, and morphologically, the conclusion is reached that carcinoma cells resemble the cells of other growing tissues. Like normal tissues, they consist of a growth fraction (stem cells and their partially differentiated progeny) and a postmitotic differentiated fraction. The latter cells are fewer in number in cancerous tissue than in normal tissue. The attributes of neoplasms that distinguish them from normal tissues appear to be largely quantitative rather than qualitative, and the neoplastic phenotype is difficult to define because most of its attributes are ones that were properties of normal cells at some time during development.

The concept of carcinoma that we propose is that carcinomas are caricatures of the process of tissue renewal. The origin of neoplastic stem cells from normal stem cells has been discussed. Support for the idea that benign tumors develop by the same mechanism as malignant ones, but from better differentiated cells than those giving rise to malignant cells, has been presented. Carcinogenesis has been considered primarily from chemical and viral standpoints, and as more and more information is accumulated the mechanism of viral carcinogenesis and transformation appears to be a problem of gene regulation and cell differentiation. Researchers in chemical carcinogenesis favor the idea that the mechanism is somatic mutation.

The equation requiring molecular explanation is:

normal stem cell + chemical, viral, or physical carcinogen → initiated but latent neoplastic stem cells.

Initiation is known to occur rapidly, but the mechanism responsible for the stability and heritability of the neoplastic process is still unknown. Stable and heritable changes may occur from alterations in the genome, through the loss or gain of chromosomes or by mutation, or apart from genomic alteration as the result of differentiation. Boveri (1929) discounted differentiation in favor of mutation because differentiation appeared to be a rapid process compared to carcinogenesis. In addition, tumors appear undifferentiated, and this diverted attention from differentiation to somatic mutation.

Chromosomal alterations as the basis for carcinogenesis were reviewed in Chapter Eight. The vast majority of human and many animal cancer cells are aneuploid. Individuals with the chromosomal breakage syndromes have a much higher than expected incidence of cancer and their non-neoplastic cells, when exposed to oncogenic viruses in cell culture, are transformed with greater efficiency than are cells derived from other individuals. Cells derived from the elderly have increased incidence of chromosomal abnormalities, and elderly persons are the major population at risk for cancer. Chromosomal abnormalities have been described in the white blood cell precursors of some individuals who subsequently developed clinical leukemia. Many carcinogenic agents cause a variety of chromosomal changes *in vitro*.

Most investigators, however, now feel that the karyotypic changes seen in neoplasms are the result of carcinogenesis instead of its cause. When karyotypic changes are present in human cancers (with the exception of the Philadelphia chromosome of chronic myelogenous leukemia) (Nowell, 1965), they are highly variable in the same kind of tumor. Even within an individual tumor, the number, size, and shape of chromosomes may vary widely from cell to cell, although marker chromosomes are sometimes present. These abnormalities change with time, and in general, there is some correlation between degree of karyotypic abnormality and malignant behavior (Defendi and Lehman, 1965).

Transformation *in vitro* would seem to be an ideal situation for studying the chronologic and pathogenetic relationships between chromosomal abnormalities and carcinogenesis. Transformed cells frequently show karyotypic abnormalities, but the relationship, if any, between such abnormalities and oncogenesis has not been discovered. In a study of spontaneous transformation of rat cells *in vitro* no chromosomal alterations that characterized the neoplastic as compared with the non-neoplastic cell populations could be found. There was no evidence that neoplastic conversion preceded, occurred concomitantly with, or followed soon after chromosomal alteration. It was concluded that the result did not eliminate the possibility of chromosomal change as a mechanism in carcinogenesis, but neither did it provide any supporting evidence for the concept.

Thus, the evidence suggests that karyotypic changes are not a necessary feature of initiation. Rather, as a result of initiation, cells appear to become karyotypically unstable, giving rise to progeny with a wide spectrum of chromosomal abnormalities. Some of these chromosomal abnormalities may confer advantages that can be selected and result in an increasingly malignant phenotype. In this way changes in karyotype could play a role in progression. It should be noted that since most karyotypic analyses were done before the advent of more refined techniques, such as chromosomal banding, the possibility remains that subtle changes in chromosomes do play a role in carcinogenesis.

Somatic mutation as a mechanism for carcinogenesis has stimulated much research (Haddow, 1944; Curtis, 1965). It is a simple explanation for the stability and heritability of the neoplastic state, and it has been coupled with the notion of dedifferentiation to account for the undifferentiated appearance of tumors. The discovery that carcinogens react with undifferentiated stem cells, and that the resultant malignant stem cells are as morphologically differentiated as the normal stem cell, does not disprove the idea of dedifferentiation; it merely bypasses it. Obviously, mutation could involve a stem cell.

It is known that chemical carcinogens can cause structural alterations in DNA (Arcos and Argus, 1968; Miller and Miller, 1971). AAF, for example, reacts predominantly with the guanosine of nucleic acids (see p. 91). This could result in alteration of a codon and might result in an altered gene product, or in its deletion. If the DNA damaged by the carcinogen is not repaired, and if the effect is not lethal, the phenotypic alteration is stable and heritable, *i.e.*, a mutation has occurred. The DNA repair mechanism itself might be at fault, allowing the accumulation of genomic damage (Lieberman *et al.*, 1971). This is thought to be the basis for the high incidence of cancer in xeroderma pigmentosum, an inherited disorder of DNA repair.

Although the concept of somatic mutation as the basis for carcinogenesis is attractive, the evidence to support it is circumstantial. The demonstration that many carcinogens are mutagenic does not prove that mutagenesis is the mechanism for carcinogenesis. As mentioned, a mutant gene might produce a new or altered gene product. With the possible exception of new surface antigens detected in the tumors of inbred animals treated with chemical carcinogens no such altered molecules have been found to date. The phenotypic changes in tumors have been quantitative instead of qualitative, leading to the postulate that mutation caused diminution or excessive production of a normal gene product that then resulted in initiation. This might point to mutation in a control operon. Since so many aspects of normal cell function are altered, there would have to be either an amplification of the effect of a single specific mutation or multiple and possibly sequential mutations.

As for specific loci, there are some who argue that there may be loci for cell growth or differentiation and that mutation in one of these loci results in neoplasms. This idea fits nicely with some of the postulates that the altered growth rate of cells might result in the phenotypic changes characteristic of progression. This is an unlikely mechanism for carcinogenesis, however, because initiated cells grow extremely slowly during the latent period. On the other hand, a mutation in a control operon could change the expression of many genes with multiple effects. It is even conceivable that a gene product altered quantitatively or qualitatively could change the cytoplasmic control, resulting in multiple and major changes in gene expression. This notion combines the mechanisms of mutation and differentiation.

There is a high correlation between carcinogenicity and mutagenicity of chemicals (McCann *et al.*, 1975). In almost all instances, however, mutagenicity has been demonstrated in drosophila or bacteria, systems that may or may not be relevant to mammalian cells. Those who believe that the situation is relevant suggest that testing for bacterial mutagenesis be used as a screening procedure in order to eliminate potentially carcinogenic chemicals from the environment (Ames *et al.*, 1973). The practicality of this proposal for eliminating carcinogens has been disputed, but it would be nice to be rid of mutagens. It should be noted that in studying the effects of chemical carcinogens on cells *in vitro*, Mondal and Heidelberger (1970) demonstrated frequencies of malignant transformation that far exceeded those of any known mutations.

Irrespective of the argument about whether or not all mutagens are carcinogens, or vice versa, it is acknowledged that most chemicals must be metabolized to a highly reactive state to be carcinogenic. These chemicals must traverse the plasma membranes, cytoplasm, nuclear membranes, and nuclear sap to reach chromosomes. It seems unlikely that they should bypass proteins or RNA or lipids or other cell constituents to reach the nucleus in an uncombined form. Rather, like steroids, they might be expected to react with a carrier protein and reach the nucleus in that manner.

Sani *et al.* (1974) have demonstrated conjugation of aminoazo dyes to a liver cytoplasmic protein to form an azoprotein. The significance of this interesting observation is not yet known, although it is known that the azoprotein dissociates into azo dyes and protein (Ketterer *et al.*, 1975). This might suggest a transport function for the protein (Sariff *et al.*, 1976).

Steroids combine with cytoplasmic carrier peptides and influence differentiation during development. Under certain circumstances in the adult, steroids are involved in carcinogenesis, whether as inducing or promoting agents is not known. If inducing, is it reasonable to assume that in the embryo steroids influence the genome and in the adult change the structure of the genome? Since it has been shown in nuclear transplant experiments that cytoplasm controls genomic expression (p. 130), it is not too improbable to suppose that the interaction of carcinogen with cytoplasmic components might upset nuclear cytoplasmic relationships affecting change in genomic expression. The specificity and degree of alteration of cytoplasmic constituents that would be required to develop the new levels of control of proliferation and differentiation to account for neoplasms are speculative. The numerous examples of derepression of fetal antigens in carcinogenesis imply that aspects of carcinogenesis need not require structural changes in DNA.

Considerable thought has gone into the role of protein as a target in carcinogenesis (Potter, 1964). Pitot and Heidelberger (1963), stimulated by the idea of feedback control circuits proposed by Jacob and Monod (1970), suggested a protein represser of gene activity produced by interaction of protein and carcinogen.

Evidence that would suggest an interaction between chemicals and viruses in the cause of cancer has been accumulating. Chemical carcinogens may activate oncornaviruses in cells not previously producing viruses. Treatment with chemical carcinogens may allow transformation of cells by viruses in instances where no infection normally occurs (Freeman *et al.*, 1973). Also, chemical carcinogens may enhance the efficiency of transformation of cells by adenoviruses (Casto, Piecznski, and DiPaolo, 1974). It is well known that x-ray-induced leukemias in certain strains of mice contain a filterable virus capable of causing the identical leukemia when injected into suitable hosts (Kaplan, 1972). Physical agents are known to effect change by either direct hit or by ionization of water with secondary chemical interactions. It should be clear from the above discussion that the critical intracellular target with which carcinogens interact has not been delineated.

We questioned the concept of somatic mutation when we made the observation that benign cells differentiated from malignant stem cells of teratocarcinomas (Kleinsmith and Pierce, 1964) and squamous cell carcinomas (Pierce and Wallace, 1971). Obviously, reverse mutation could not account for the phenomenon, and we proposed an alternative to the concept of somatic mutation as the cause of neoplasms

(Pierce, 1967). The alternative that was compatible with normal development and neoplasia proposes that neoplasms are postembryonic differentiations (Pierce, 1967, 1974).

In our concept, the target in carcinogenesis is an undifferentiated stem cell normally responsible for tissue renewal, instead of a differentiated cell, which is induced to dedifferentiate. As a result of the alteration of the controls of its genome, the normal stem cell differentiates into a malignant stem cell. The malignant stem cells give rise to a tissue in which some phenotypic characteristics differ from those of the progenitor tissue. The concept of differentiation as the mechanism of neoplasia accounts for the stability and heritability of the neoplastic process previously attributed to somatic mutation.

In considering the developmental concept of carcinogenesis, it should be stressed again that all *normal* tissues arise by differentiation and are maintained by a group of determined stem cells that proliferates and differentiates under regulatory controls, thereby maintaining the status quo of the mature organism. Since cancer is a tissue, it would seem reasonable that its mode of origin and mechanism of regulation and development might be comparable to that of normal tissues. The preponderance of stem cells in the cancer and the clinical impact of the tumors have tended to misdirect us from this very logical conclusion.

The essential aspects of a differentiation require the interaction of an environmental stimulus or inductor with competent precursor cells. A new tissue with unique, stable, and heritable properties distinguishable from the progenitor results from this reaction (Grobstein, 1959). A neoplasm obviously fulfills these definitional requirements: Numerous environmental stimuli have been identified, competent undifferentiated precursor cells have been identified as the cell of origin; and a new tissue is produced with unique, stable, and heritable properties distinguishable from the progenitor. We have not recognized these attributes because the undifferentiated appearance of tumors has misled us.

If the concept that the mechanism of carcinogenesis is similar to that of differentiation is true, then it is axiomatic that all of the genetic information necessary for expression of the malignant phenotype be encoded in the normal precursor cell (Pierce, 1974). The only way to demonstrate what information is contained in the genome is to identify situations in which it is expressed. Important attributes of the neoplastic phenotype that would have to be explained in this manner are the capacity for neoplastic cells to proliferate at the expense of differentiation, to usurp the logistics of the host and the ability to invade and metastasize.

Normal cells exhibit each of these properties at some time during development. For example, the fertilized egg has a marked capacity for proliferation with little overt evidence of differentiation until a sizable number of cells has been achieved. As for usurpation of the logistics of the host, the fetus has preferential treatment even at the expense of the maternal organism. Invasion, with all of the aggression that the term implies, is known to developmental biologists as *migration*. For example, neural crest cells migrate from the primordial neural tube and invade the surrounding tissue (Weston, 1971). Ultimately, they are disseminated throughout the body. As for metastasis, cells of the reticuloendothelial system arrive at their anatomical locales by either vascular or lymphatic dissemination; and germ cells in birds, which develop in the yolk sac, invade the circulation and "metastasize" to the primitive gonad. Thus, it can be concluded that the summation of characteristics that the diagnostician uses as criteria for malignancy are all properties encoded in the normal genome. As a result of carcinogenesis, these parts of the genome repressed during normal development are derepressed. These effects are all stable and heritable (Pierce and Johnson, 1971). Thus, neoplastic cells appear to be out of context temporally and the malignant phenotype is inappropriate for longevity of the host, but it is worth stressing again that this is incidental to the mechanism of neoplastic development. We survive as a species because the lethal effects of most neoplasms occur after the reproductive period.

Some neoplasms synthesize molecules not normally produced by the adult tissue of origin. For example, some bronchogenic carcinomas synthesize human gonadotropin (HCG) (Lipsett, 1968). Presumably, the gene loci responsible for the production of these hormones were derepressed during the process of carcinogenesis. It would be an extreme stretch of the imagination to believe that the cells of the carcinoma had undergone a systematic reversion and reacquisition of potential with later differentiation into trophoblast (dedifferentiation and redifferentiation). The cells of these lung tumors do not look like syncytiotrophoblast and lack the other functions of trophoblast, and it would seem more reasonable to expect that in carcinogenesis the gene loci of lung cells governing HCG synthesis were derepressed. As is well known, activation and repression of the genome are the mechanisms of differentiation; in tumors, growth and differentiation simply are not integrated with the needs of the organism and are often unresponsive to usual controls. In this sense they are aberrant.

Hormone production by tumors of non-endocrine tissue is easy to recognize because of the biological potency of the agents involved. There are other examples of inappropriate gene activation by chemical carcinogens, including antigens and enzymes, many of which were originally thought to be *tumor specific* and now are known to be characteristics of rapidly growing non-neoplastic tissues (Gold and Freedman, 1975; Coggin and Anderson, 1972, 1974).

Recently, Comings (1973) also proposed a theory of carcinogenesis predicated upon the hypothesis that all cells contain the genetic information for the cancer phenotype. As discussed above, this information is an intrinsic component of the genome of normal cells and presumably plays an important role during development. With maturity these loci are normally repressed by regulatory genes. Activation of the cancer loci would occur if both members of a pair of regulatory genes were eliminated, a feat that could also be accomplished through a mutation affecting both members of the pair of regulatory genes. Tumors that are inherited as autosomal dominants (retinoblastoma) might represent a situation in which a cell inherits a defective regulatory gene and the occurrence of a tumor is dependent upon a mutation of the second member of the pair of regulatory genes (Knudson, 1971). In this theory, oncogenic viruses must have been derived from the genome of the cell.

Although a mutational event is emphasized in Comings' thinking, epigenetic events could function equally well. Once again, if carcinogenesis is simply a stable, heritable alteration in phenotype (differentiation), and it is unnecessary to invoke mutation to account for other differentiations, why is it necessary to invoke mutation as a mechanism in cancer? It would seem that epigenetic events are as likely to determine the effects of cancer-controlling genes as they are to determine the effects of other controller genes.

Reconsider the experiments with platyfish (Anders, 1967) (see p. 116). Selective crossing of these fish with another species resulted in the production of malignant melanomas in the hybrids. The genomic information for melanomas was present in the animals, but it was not expressed until a suitable cellular environment was experimentally produced. A mutation in the germ line of some ancestor of these animals could have been segregated by selective breeding. If this occurred, the expression of the melanoma phenotype in succeeding generations and late in life is obviously a matter of control of gene expression.

Various explanations are available for the introduction of information for neoplasia into the genome. The AKR leukemia is an example (Rowe, 1973). This tumor is caused by a C-type virus that has a copy of its RNA genome inserted in the cellular genomes by reverse transcriptase. It turns out that once incorporated in the genome of germ cells, the viral information is passed from parents to offspring according to Mendelian genetics. Thus, some animals get leukemia; others do not. Whether the AKR mouse develops a tumor or not depends upon the expression of the viral genome. Control of gene expression is presumed to be the mechanism of differentiation. A differentiated tissue has stable, heritable properties because of controls of genome rather than structural changes in DNA.

In summary, normal stem cells contain the necessary information for expression of the malignant phenotype. Possibly a mutation in an ancestor is passed according to Mendelian genetics and when it is suitably paired, it sets the genetic stage for neoplasia. Whether or not a tumor develops will depend upon the environment of the cells. Similarly, a virus may create an oncogene by putting information into the germ line with a reverse transcriptase and set the stage for development of a neoplasm. In each of these instances, the development of a tumor in a particular tissue, and not in others, is dependent upon control of the genome.

In addition to cancer genes, chemical carcinogens, physical carcinogens, and viruses can cause cancer locally by either causing a somatic mutation or by altering control of a genome. If the latter is true and there is no evidence to the contrary, then we must study the mechanisms of gene control, the social interaction of cells, the controls of cell division, and differentiation as new approaches to therapy of cancer.

Controls of gene expression

Much of what is known about controls of gene expression has come from studies of nuclear transplantation and cell hybridization. Since these are complex subjects in themselves, it is necessary to review them as topics before conclusions can be drawn from them regarding cancer.

Nuclear transplantation

The initial thrust of the nuclear transplantation experiments of Briggs and King (1952) and Gurdon (1963) had to do with whether or not irreversible changes occurred in the genome during differentiation. It was conceivable that the genome remained intact during differentiation with selective repression or derepression of its appropriate parts, or that parts of it might be deleted or changed.

Nuclear transplantation was envisioned by Briggs and King (1952) as a direct means of solving this problem. If the nucleus of a differentiated cell had lost much of its genome in the process of differentiation, when transplanted into an undifferentiated cytoplasm, it would direct the production of its specific differentiated molecules and cells. If, however, the genome were intact, and if all other things were equal, one might expect development in accord with the state of differentiation of the recipient cytoplasm. This, of course, is predicated upon the notion that the cytoplasm controls genomic expression. In the "whereases" and "ifs" reside major and still unresolved problems, but as the experiments turned out, cytoplasm was capable of controlling gene expression and the genome of the differentiated cell appeared to be intact.

Briggs and King (1953) overcame the tremendous technological problems in the successful transplantation of nuclei of amphibians. Briefly, the oocyte must be activated, its nucleus removed, and a viable nucleus from a differentiated cell transplanted into the prepared oocyte. Exposure of the donor nucleus to culture medium results in its destruction. Differentiated cells are small, and the technological difficulties involved in breaking the cell membrane (an absolute requirement if the transfer is to be successful), protecting the nucleus from tissue culture media, and placing it within the oocyte are most exacting. Because of the disparity in size between the oocyte and the differentiated cell, it is impossible to transfer the nucleus of the oocyte into a differentiated cytoplasm. Thus, there are no *back controls* for any of the data.

Briggs and King transplanted nuclei from *Rana pipiens* and observed development to adult stages. Many of the recipients were abnormal or arrested at the first mitosis or at blastula or gastrula stages. Notable was the observation that ova with transplanted nuclei from differentiated cells did not develop into a mass of differentiated cells of donor type. Equally interesting was the observation that the older the donor embryonic cells, the fewer the nuclei that could promote normal development in test eggs. This led to the conclusion that a progressive nuclear change occurred during embryonic differentiation. Gurdon (1963), working with larvae of *Xenopus laevus*, the South African clawed-toad, found that in 7% of transplants development proceeded in an apparently normal manner and gave rise to adult animals. The donor nuclei were taken from gut cells and Gordon postulated that the genome of differentiated cells was intact. Similarly, Simnett (1964) obtained normal development with 38% of nuclei transplanted from cells from the brain rudiment.

The discrepancy between the data of the various transplanters is more apparent than real (McKinnell, 1972). Each has observed more restriction in development when nuclei from highly differentiated cells are transplanted. Only a few of them developed normally and 50% of all transplanted nuclei failed to divide even once. DiBerardino and King (1967) and DiBerardino and Hoffner (1971) noted that successful nuclear transplants had normal chromosomes, but those of arrested transplants were frequently aneuploid. It turned out that the transplanted nucleus was induced to undergo mitosis shortly after transplantation. The swelling of these nuclei was not uniformly regular. Condensed areas of chromatin were unable to decondense, and the first mitosis of these cells resulted in lethal karyotypic abnormalities. DiBerardino and King (1967) suggest that the donor nucleus must be in an appropriate premitotic phase to enter successfully the cleavage cycle of the recipient egg's cytoplasm. When it is, there is normal development, which supports the idea that the genome of the differentiated cell is intact.

Consider for a moment other implications from these experiments. The cytoplasm of the egg must control genetic expression because an embryo developed from eggs with transplanted nuclei of differentiated cells. If genome controlled cytoplasm there would have been development of a mass of cells of donor type. This is in accord with the notions of Graham (1966), and with the observation that cytoplasm also controls the mitotic cycle (see p. 74). In this regard, enucleated cytoplasm can divide at least once (Prescott, 1976). Ursprung and Markert (1963) found that injections of protein into egg cytoplasm always resulted in arrested development.

Of importance in understanding neoplasia as a differentiation are experiments performed by King and McKinnell (1960) and King and DiBerardino (1965) involving the transplantation of nuclei from renal cell adenocarcinomas of frog into the cytoplasm of activated, enucleated eggs. This tumor is believed to be caused by integration of DNA of a herpesvirus with the DNA of the cell. If this viral information had irrevocably altered the genome of the cell in favor of neoplasia, one would anticipate that the transfer of such a nucleus into an egg would result in the rapid proliferation of a group of undifferentiated renal adenocarcinoma cells. This proliferation would continue until the yolk of the egg was consumed, and the organism and the tumor cells would succumb, a situation resembling the natural history of carcinoma in mammals. However, this sequence of events did not occur. What was observed was the development of abnormal larvae with apparently normal organogenesis, including the formation of eyes, heart, and other organs. Although the larvae eventually died, it was clear that they were not neoplastic. Possibly their demise was the result of virus production in the cells.

The origin of the nuclei in these experiments has been of concern. Were they derived from the cancer cells or from stromal cells? Subsequent experiments by McKinnell, Steven, and Labat (1976) support the conclusion that adenocarcinoma cells were the source of nuclei, but it is impossible to prove that a particular nucleus came from a renal adenocarcinoma cell or from a stromal cell. If the transplanted nuclei originated in cancer cells, two important deductions can be made. Not only did the cytoplasm of a recipient oocyte control neoplastic expression, it also controlled neoplastic expression in a nucleus with an integrated viral oncogenic genome.

There are other situations in which viral-induced malignant tumor cells differentiate to a benign state. Functional, well differentiated, postmitotic senescent cells are produced in a mammary adenocarcinoma of the mouse (Wylie, Nakane, and Pierce, 1973). This suggests that either the mammary tumor virus is excluded from the cell during the process of differentiation or the expression of its genome is somehow controlled. In view of Rowe's studies on AKR leukemias (1973), it would seem that the viral genomes are controlled.

If the cytoplasm controls gene expression, and if it is agreed that the controls of differentiation are sufficiently stable to account for the stability and heritability of malignant expression, then one could postulate that instead of interacting directly with DNA, chemical carcinogens might react with a nucleophilic molecule in the cytoplasm. The resultant molecule would feed back new signals into the nucleus causing repression or depression of the genomic loci evolving the changes characteristic of malignancy.

The loss of malignant attributes by neoplastic cells, which spontaneously differentiate, could result from the accumulation of differentiated products in the cytoplasm which, in turn, would feed back information into the nucleus, repressing and derepressing loci associated with the control of malignant properties. This is obviously an oversimplification of a very complex situation, but it suggests alternatives to mutation as the only mechanism of neoplasia.

Cell hybridization–cell fusion

As with nuclear transplantation, the interpretation of data from studies of cell hybridization as they relate to malignant change requires an overview. Cell hybridization as a tool for the study of somatic cell genetics had its origin in the demonstration of Barski *et al.* (1961) that high malignancy and low malignancy cell lines when mixed *in vitro* spontaneously produce hybrids that share properties of each parent. These could be isolated as clonal lines (Barski, Sorievl, and Cornefert, 1961; Barski and Cornefert, 1962). The growth rate and malignancy of these lines approximated that of the malignant parent. In other words, the malignant phenotype was not repressed by hybridization with a nonmalignant parent. Upon passage *in vivo* or *in vitro*, progressive chromosomal loss was noted.

Gershon and Sachs (1963) mixed cells from a polyoma-induced mammary tumor with a strain of L-fibroblasts *in vitro*; 26 days later approximately 1% of the cells had a new phenotype characterized by large nuclei and an altered growth rate. Morphologically the cells were intermediate in their arrangement between the extreme crisscross pattern of the polyoma-induced parent and the flat growth pattern of the L-cells. When karyotyped, the hybrids had 108 chromosomes (the sum of the parental chromosomes would have been 124) and they were as malignant as the polyoma-induced parent.

Ephrussi and his associates (Ephrussi, 1972) also studied spontaneously derived hybrids from high and low tumor strain cells. Scaletta and Ephrussi (1965) obtained a hybrid between a high cancer line with normal cells of CBA origin carrying marker chromosomes. Hybrids had a modal chromosome number of 90 and their tumorigenicity approached that of the malignant parent. Subsequently, Finch and Ephrussi (1967) fused L-strain fibroblasts to teratocarcinoma cells of strain 129 mice and found that the resulting tumors grew rapidly *in vivo*, but they lacked the ability to differentiate into the multiplicity of tissues characteristic of teratocarcinoma. These results were confirmed by Jami, Failly, and Ritz (1973).

Multinucleated giant cells are commonly found in measles and other viral diseases, an observation that led Okada to the notion that viruses might be utilized to form experimental multinucleate giant cells. Okada and Tadokoro (1962) discovered that destruction of the nucleic acid of Sendai virus did not remove the capacity of the virus to cause cell hybridization. Thus, ultraviolet light inactivated Sendai virus provided a tool for fusing cells, a technique that was successfully exploited by Harris and Watkins (1965). Four hours after exposure of cells to high doses of virus, approximately 10% of the cells had fused into multinucleated giant cells. At low doses of virus, many cells had two nuclei and, within 5 days most of these had fused into a single nucleus. Fusion of nuclei took place during mitosis.

In each study of hybridization, isolation of the hybrid line depended upon its ability to overgrow the parental lines or upon the ability of the investigator to successfully isolate the hybrid cells. Selection of hybrid cells was made practical by taking advantage of metabolic defects in parental lines. These could be corrected in the hybrid as the result of input from the other parent (Littlefield, 1964, 1966). A9 cells lack hypozanthene-guanine phospho-ribosyl transferase (HGPRT). B82 cells lack thymidine kinase. When endogenous synthesis of purines and thymidylic acid is blocked with aminopterin, these deficient cells do not survive because they cannot use hypoxanthine or thymidine. Although each parental cell is deficient in one enzyme, the hybrid contains both enzymes, survives exposure to aminopterin, and can be quickly and easily cloned (Fig. 8–1).

Of considerable importance in the interpretation of the data are the events occurring during the period before hybrid cells can be isolated. This may be as short as 1 or 2 weeks for rat-mouse hybrids or as long as 6 weeks for human-mouse hybrids. During this interval a reduction in the number of chromosomes occurs, a phenomenon agreed upon by

Fig. 8–1   The selection of hybrid cells using HAT medium. (a) TK⁻ cells are mutant cells that lack thymidine kinase (TK). This enzyme converts thymidine generated from the breakdown of DNA into deoxythymidine monophosphate (dTMP) that is subsequently utilized in DNA synthesis. A TK⁻ cell will die in the presence of aminopterin because this blocks alternative pathways for the synthesis of dTMP. A TK⁻ cell can survive treatment with azaguanine. (b) HGPRT cells are mutant cells that lack the hypoxanthineguanine phosphoribosyl transferase. This enzyme salvages adenine, guanine, or hypoxanthine from breakdown of RNA or DNA and phosphorylates these nucleic acids. The products adenosine monophosphate or guanosine monophosphate are used in the synthesis of DNA. HGPRT⁻ cells cannot incorporate the base analogues thioguanine or azaquanine and are resistant to the toxic effects of these drugs. These cells are resistant to treatment with aminopterin in the presence of thymidine. (c) HAT medium contains hypoxanthine, aminopterine, and thymidine. HGPRT⁻ and/or TK⁻ cells die in this medium. (d) Hybrids of HGPRT⁻ and TK⁻ cell lines survive in HAT medium because the hybrid contains the wild-type form of HGPRT and TK.

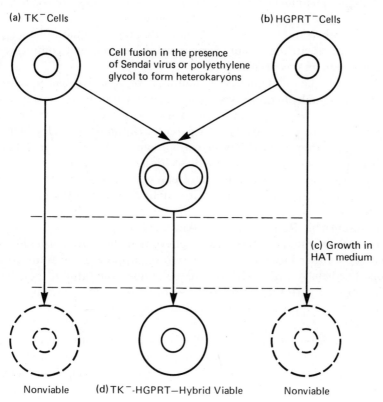

(a) TK⁻ Cells   (b) HGPRT⁻ Cells

Cell fusion in the presence of Sendai virus or polyethylene glycol to form heterokaryons

(c) Growth in HAT medium

Nonviable   (d) TK⁻-HGPRT—Hybrid Viable   Nonviable

all investigators. This must be taken into account in studies of malignancy.

Malignancy has been analyzed in hybrids derived by hybridizing highly malignant cells with cell lines of low malignancy or diploid fibroblasts from primary cultures (Harris, 1971; Klein and Harris, 1972; Bregula, Klein, and Harris, 1971; Weiner *et al.*, 1972). When five malignant lines of cells, including Erhlich and polyoma transformed cells, were each hybridized with A9 or B82 derivatives of L-strain fibroblasts (low tumorigenicity), the tumorigenicity of the hybrid cells approached that of A9 or B82 cells. These hybrids contained a chromosomal complement close to the sum of the parental contributions. Tumors obtained by transplantation of the hybrids consisted of cells that had lost significant numbers of chromosomes. Similarly, when polyoma transformed cells were hybridized to A9, tumorigenicity paralleled that of the L-cells if a complete or nearly complete chromosomal complement was present (Klein and Harris, 1972). All hybrid cells were karyotypically unstable, however, and as chromosomes were lost, the phenotype approached that of the malignant parent.

Silagi (1967) obtained spontaneous hybrids from a line of pigmented malignant melanoma cells and A9 strain fibroblasts. Some of the hybrid lines produced tumors on transplantation when a large inoculum was used, but others did not. Silagi concluded that the malignant phenotype was dominant. In the tumors karyotyped, a near normal complement of chromosomes was present.

Thus, whether malignancy is dominant or recessive has not been settled, although there are strong opinions on the matter. Klein and Harris (1972), using viral-induced hybrids, concluded that as long as hybrids contain nearly the sum of parental chromosomes, they behave in benign fashion. With loss of chromosomes, tumorigenicity of the hybrid increases. In contrast to these ideas, spontaneously occurring hybrids containing nearly the sum of parental chromosomes have been observed to have high tumorigenicity mimicking that of the malignant parent.

Hitosumachi, Rabinowitz, and Sachs (1971), on the basis of their work with spontaneous revertants, believe that there are chromosomes that carry either loci for transformation or suppression of transformation. The behavior of a cell would depend upon the ratio of these loci. This idea fits with the notion that loss of chromosomes carrying information that suppresses transformation results in a malignant hybrid cell.

Wiener, Klein, and Harris (1974) also proposed that the specific lesion in malignancy is some type of genetic loss that is restored when the malignant cell is fused to a normal cell. As long as the chromosomes responsible for suppression are present, the hybrid cell will behave in benign fashion. When these chromosomes are lost, the cells express the malignant phenotype once again. They propose that carcinogens operate not by converting normal cells to malignant cells but by generating genetic variation to allow malignant variants (presumably with genetic loss) to occur. Appropriate environmental conditions would result in selection of malignant cells and the development of a tumor.

There are suggestions that hybrid cells are produced *in vivo*. Stone, Friedman, and Fregin (1964) have observed evidences of cell fusion in non-identical twin cattle. The calves share a common placenta and exchange hematopoietic cells *in utero*. Thus, each twin contains hematopoietic elements from its sib and has immunological tolerance to the foreign cells. By the time the calves are 8 years old a third population of cells appears that seems to be a hybrid of the two parental lines. It is not known if these animals have a high incidence of hematopoietic malignancies. Although it is conceivable that malignancy might be the result of spontaneous cell hybridization *in vivo*, with loss of chromosomes leading to the aneuploid state, the demonstration that some malignant tumors and all benign tumors are euploid casts doubt upon this as the mechanism for oncogenesis. In general, the results of hybridization studies have thus far been disappointing and have added little to the understanding of the mechanism of carcinogenesis.

Gene expression

It can be postulated from the experiments in nuclear transplantation that cytoplasm controls genomic expression, but the manner of this control is not understood. Biochemists have searched in the nucleus for agents capable of interacting with, and repressing, DNA. The histones have received much attention, and more recently nonhistone, chromatin-associated proteins have been extensively studied. The molecules that interact with DNA may be compared to switches capable of turning genes off and on under the control of cytoplasm (Stein, Spelsberg, and Kleinsmith, 1974; Monahan and Hall, 1974).

The histones are a group of small (from 10,000 to 25,000 molecular weight) basic proteins found in the chromosomes of eukaryotic organisms and are subclassified according to their lysine or arginine contents. They probably play a structural role in chromosomes, and in *in vitro* systems they restrict the template activity of DNA for RNA synthesis. Histones modified by phosphorylation or acetylation have been described, but these modifications seem insufficient to confer a primary role on histones for gene control.

The nonhistone, acidic, chromatin-associated proteins are heterogeneous (from 5,000 to 100,000 molecular weight) and rich in glutamic and aspartic acids. Experiments in which chromatin and these proteins are reassociated result in transcription that can be quantitatively as well as qualitatively changed by varying the reactants. The possible roles of these proteins in control of gene expression during differentiation have been suggested by the presence of specific nonhistone, acidic protein associated with the development of specific tissues.

Since hormones are capable of evoking specific differentiations, and are themselves often well characterized chemically, their mode of action has been extensively studied. For example, estrogen is known to bind to receptor proteins. Jensen *et al.* (1972) have found a cytoplasmic 4S estrogen binding complex in the uterus that must be transformed to a 5S complex before nuclear binding and stimulation of RNA synthesis can occur. The stimulation of RNA synthesis by the 5S estrogen complex is specific for uterine nuclei. The precise mechanism by which this reaction takes place is unknown, but because of the tremendous number of estrogen receptor molecules transformed and then trapped in the nucleus, it has been speculated that these complexes act on RNA polymerase. However, there are different classes of binding sites for steroid protein complexes on chromatin. It has been postulated that only a few of these complexes are active in altering the transcription of specific genes.

The chick oviduct has been useful in the study of the actions of estrogen and progesterone. Estrogen causes DNA synthesis, cell division, and cytodifferentiation of three types of cells from mucosa of the primitive oviduct. There is a marked increase in the production of a specific protein, ovalbumin. O'Malley and Schrader (1976) have found that estrogen causes qualitative and quantitative changes in RNA transcription that can be blocked by actinomycin D. The level of mRNA specifically coding for ovalbumin has been measured by isolating the mRNA from the stimulated oviduct and assaying the translation of ovalbumin in cell-free, protein-synthesizing systems. After mRNA for ovalbumin was purified, a DNA copy was synthesized *in vitro* by using an RNA-directed DNA polymerase and labeled precursors. This labeled DNA was used as a probe in nucleic acid hybridization reactions with chick oviduct DNA. Kinetics of this hybridization indicates that there is only one gene copy for albumin per haploid genome. Thus, selective gene amplification, a process by which genes for ribosomal RNA are increased in amphibian oocytes, is not the explanation for the increased ovalbumin production in response to estrogen stimulation. Since there is only one gene copy for ovalbumin and the rate of mRNA for ovalbumin markedly increases with estrogen stimulation, there must be transcriptional control. There is also an increase in ribosomal RNA and transfer RNA, in addition to a marked increase in ovalbumin synthesis in response to estrogen.

Progesterone causes cells in the chick oviduct to synthesize the protein avidin. Cytoplasmic receptor proteins combine with progesterone to form complexes that bind to nuclei or chromatin. Binding of these progesterone protein complexes to chromatin requires the presence of specific nonhistone proteins. The exact mode by which transcription is subsequently controlled is unknown; it may be by direct action on the chromatin template or on RNA polymerase.

Baxter *et al.* (1972) have studied the production of an enzyme, tyrosine aminotransferase (TAT), in cultured rat hepatoma cells in order to understand control of gene expression. The enzyme is induced with a steroid that crosses the cell membrane and binds to specific protein receptors; these migrate to the nucleus. Here the steroids probably interact with chromatin to stimulate the synthesis of mRNA specific for TAT. When the steroid is removed from the system, the steroid receptor leaves the nucleus and the rate of synthesis of TAT returns to pre-induction levels.

Translational controls could work through alterations in any of the components involved in the translational mechanism, including tRNA, initiation and elongation factors, ribosomes, and mRNA. Analysis of *in vitro* protein synthesizing systems has shown that there can be selective translation of messenger RNAs depending on the ionic environment, tRNAs, and ribosomes, but it has been difficult to determine if these components can be limiting within cells.

There are other controls in eukaryotic cells that appear to involve the translational process. For example, enzyme levels may change markedly in the absence of RNA synthesis. An example is the threefold increase in serine dehydratase in the liver after administration of amino acids, even while new RNA synthesis is blocked. Such studies have shown that messenger RNAs can have variable lifetimes. The estimated lifetime of mRNA for serine dehydratase in rat liver is from 6 to 8 hours, while in certain hepatomas it can be greater than 2 weeks. Pitot *et al.* (1974) believe that the phenotypic characteristics that make tumors different from their tissue of origin could be controlled by the variable stability of the templates for protein synthesis. The control of the rate of turnover of mRNA species is not understood.

Control of gene expression has also been studied by using DNA tumor viruses. Using molecular hybridization it has been found that most of the viral RNA extracted from transformed cells is made up of what is termed *early gene* information. RNA extracted from lytically infected cells contains not only early gene RNA, but also *late gene* RNA. Transcription of the SV_{40} genome appears to involve strand switching. Early gene functions are transcribed in one direction on one strand of DNA and the late functions are transcribed from the anti-parallel complementary DNA (Khoury, Bryne, and Martin, 1972; Sambrook, Sharp, and Keller, 1972).

The relationship between the control of RNA produced and transformation is not fully understood. The size of RNA, complementary to the viral DNA, in the nucleus and its transport to the cytoplasm has been studied because this might be another type of control mechanism. It appears that the nuclear RNA is larger than that found on polysomes and must be processed before it can be transported to the cytoplasm (Lindberg and Darnell, 1970). Fusion of a permissive cell (even an enucleated permissive cell) with a nonpermissive transformed cell that has integrated the SV_{40} genome allows for productive infection (Gerber, 1966; Watkins and Dulbecco, 1967; Koprowski, Jensen, and Steplewski, 1967; Croce and Koprowski, 1973). This suggests that there is a cytoplasmic factor necessary for productive infection. The mode of action of this postulated factor is unknown.

RNA viruses have also been studied with regard to the control of gene expression. The discovery of posttranslational processing of the proteins of poliovirus suggests that enzymes capable of tailoring large proteins are important in synthesizing functional proteins (Korant, 1975). The presence or absence of proteolytic enzymes may influence the proteins and subsequent phenotype of a cell.

There are also stringent requirements that determine whether RNA tumor viruses replicate or transform the cell. The Rous sarcoma virus (RSV) infection of mammalian cells is markedly different from that seen with RSV infection of chicken cells (Bishop and Varmus, 1975). The chicken cells both transform and produce infectious virus. However, mammalian cells that contain RNA with viral sequences are not capable of producing infectious virus. It is not yet clear whether or not complete transcription is occurring. Because fusion of the mammalian RSV-transformed cells with chicken cells or the infection of chicken cells with the DNA from RSV transformed rat cells allows production of infectious virus, it is hypothesized that the cytoplasm of the mammalian cell lacks the ability to translate the RSV.

The production of RNA tumor viruses can be either induced in cells not previously producing virus or enhanced several hundredfold in cells producing virus particles. The induction occurs with a variety of chemicals, including halogenated pyrimidines, chemical carcinogens, protein synthesis inhibitors, and steroids. Induction is thought to involve the derepression of transcriptional controls. The role of virus induction in carcinogenesis is being studied.

From the brief description of these models, the complexities of control of the genome should be obvious, and it should be equally obvious that through understanding these mechanisms will come the understanding of differentiation in general and of carcinogenesis in particular. At this time, much of the available data deals with the identification of the switches acting at the levels of transcription and translation. The next step is to determine the means by which the switches are turned on or off and, ultimately, the overall means by which the system is integrated.

In addition to the controls that act at the level of the gene, there appears to be at least one other level of genetic control; its existence is suggested by nuclear transplantation experiments and studies of heterochromatin. In nuclear transplants the transplanted nucleus is induced to undergo division immediately, in response to a cytoplasmic signal. In unsuccessful transplants, unequal swelling of chromatin is followed by faulty mitosis and karyotypic anomalies, many of which are lethal (DiBerardino and Hoffner, 1971; Gurdon, 1974). The unequal swelling suggests different kinds of packaging of DNA along the chromosome and possibly among chromosomes. This could also serve as a control. Further evidence for this idea comes from studies of heterochromatin. Early in the embryonic development in the female, one of the X-chromosomes in each cell is permanently repressed (Lyon, 1968). The repressed X consists of densely aggregated or clumped chromatin that is recognized as heterochromatin. The portion of the genome that is heterochromatic is greatest in differentiated cells and it is most marked in erythroid cells prior to extrusion of the nucleus. Heterochromatin is thus an economical way for the cell to package genetic information no longer appropriate for current function. The active parts of the genome are probably controlled at the level of the gene by finer and more labile controls, such as those described by Jacob and Monod (1970). That both heterochromatin and the operon circuits are under cytoplasmic control is evidenced by the ability of the nucleus from a chick erythrocyte to be stimulated to synthesize DNA after fusion with a fibroblast (Harris, 1967). Chemical carcinogen might act by altering the control of heterochromatin, activating large segments of genome, including those controlling the many aspects of the malignant phenotype. This mechanism obviates the need for amplification of the signal to account for the many phenotypic changes in cancer. Obviously, alteration of control of a single gene could produce changes in a gene product and also alter cytoplasmic control of the genome.

These notions raise more questions than they answer. What are the changes in gene products that lead to malignant expression? How are these stimuli mediated in the nucleus? With what do chemical carcinogens interact in the cytoplasm to cause neoplasia? Is it conceivable that each of these factors mediates its effect by alterations in control of growth? Is there a chromosome or a portion of a chromosome that is specifically associated with cell motility, invasion, metastasis, and certain fetal antigens, all of which could be activated as a unit?

Apparently during normal development there is a programmed repression of cellular attributes that, if present in the adult, are described as neoplastic attributes or the neoplastic phenotype. Conversely, the accumulation of differentiated products in the cytoplasm of embryonal carcinoma cells in most instances leads to loss of neoplastic attributes, presumably as a result of cytoplasmic control of genomic expression. This fits in nicely with what is known of normal cytoplasmic controls of gene expression.

It has long been held that there is a need for inductive stimuli if a differentiation is to proceed. Holtzer (1970) has suggested that differentiation may be preprogrammed within the cell, and given the appropriate environment, it will proceed according to plan. Holtzer postulates two kinds of mitosis in development: One merely increases the number of cells and the other causes diversification as a result of uneven division of cytoplasm. These postulates for differentiation, either unequal cell division or environmental stimulation, are not mutually exclusive, and both require appropriate environmental conditions for the differentiation to occur.

Growth controls–repressors

The controlling effects of hormones and vitamins and the modulating influences of the intimate cellular environment have been discussed in previous sections. Discussion here will center around a variety of inhibitors and stimulators of growth that are now being isolated from various tissues.

Chalones

In 1957 Weiss and Kavanaugh described regulation of growth in mathematical terms. Growth was considered the net gain of organic mass over that lost. The mass consisted of two components: the generative mass endowed with reproductive activity and the differentiated mass derived from the generative but unable to reproduce. It was postulated that each tissue synthesized *templates* that catalyzed the reproductive process and were controlled in a negative feedback mechanism by *antitemplates*. Adult size would be achieved when equilibrium between incremental and decremental growth occurred, presumably as a result of balance between template and antitemplate activity.

Experimental support for this hypothesis was obtained by Bullough and Laurence (1962) while looking for agents that would promote wound healing. Instead, aqueous extracts of skin adjacent to wounds were found to contain agents that inhibited mitosis. These were named *chalones*.

The first evidence that chalones were tissue specific was derived from the observation that they inhibited mitoses in the basal layer of skin, but not in adjacent sweat glands. Subsequently, it was shown that although chalones are tissue specific, they are not species specific. Now most experiments with skin chalones are performed with material derived from the skin of hogs.

Extracts from basal cells do not inhibit basal-cell mitosis, but similar extracts prepared from keratinizing epithelium do inhibit mitotic activity in basal cells. This suggests that the differentiated cell mass exerts feedback control over proliferation of the generative mass.

Antimitotic agents have been isolated from liver, bone marrow, lymph nodes, kidney, endometrium, and lung. Again, these chalones are tissue specific but not species specific and in no instance have they been purified or their molecular structure determined. Chalones extracted in aqueous solution and concentrated by precipitation with from 70% to 80% alcohol are found to be glycoproteins with molecular weights of approximately 40,000. They are nondialyzable, trypsin labile, and denatured by boiling. It has been shown that they require cofactors (*e.g.*, epinephrine or hydrocortisone) to produce their effect (Houck and Attallah, 1975).

Multiple kinds of cancer cells have been shown to respond to their tissue-specific chalone by inhibition of mitotic activity. Bullough and Laurence (1968) isolated chalone from the rapidly dividing cells of the V-2 squamous cell carcinoma of rabbits and found high levels in the serum of tumor-bearing animals. When chalone was administered to these tumor cells, they responded by decreased mitotic activity. It was then shown that the intracellular level of chalone in the tumor cells was decreased in relation to normal cells. It was postulated that, although tumors could make the appropriate chalone, it was quickly lost from the cells and, therefore, the chalone did not inhibit mitotic activity of the tumor cells *in vivo*. Interestingly enough, there was depressed mitotic activity in the normal skin of animals bearing this transplantable tumor.

Rytömaa and Kiviniemia (1968) found that cells of a leukemia contained from one-tenth to one-fortieth of the chalone found in normal leukocytes and the tumor cells were less sensitive to the inhibitory effects of extracellular chalone. Since these cells proliferated at a slower rate than normal leukocytes, Rytömaa (1973) postulated that leukemia represents less of an enhancement of growth rate and more of delayed maturation of cells allowing for an enlarged growth pool. It is his belief that if the growth of tumors could be slowed by chalone, this would allow spontaneous differentiation to occur with ultimate cure of the disease.

It now appears that there may be more than one kind of chalone in a tissue and that these chalones may operate at different points in the mitotic cycle. For example, one chalone arrests mitosis in G_2 and another apparently arrests mitosis in G_1.

Growth controls–stimulators

A host of factors that stimulate growth of various tissues has been isolated. Already mentioned are colony-stimulating factors, which promote the growth and differentiation of leukocytes; erythropoietin, which has been briefly mentioned in relationship to the erythroleukemia of Friend; and vitamin A, which modulates the expression of squamous epithelium and in large amounts directs this phenotype to one that is mucus-secreting. Trophic hormones of the pituitary and other endocrine glands have growth-promoting or inhibitory effects upon various target organs, but less well known is the role of the submaxillary gland in the elaboration of certain growth factors. The submaxillary gland has been found to synthesize a nerve growth factor, an epidermal growth factor, and a factor that appears to play a role in lymphopoiesis.

Nerve growth factors

The effect of nerve growth factor was first observed in experiments using snake venoms and certain sarcomas, but the highest concentration has been found in the submaxillary glands of rodents. Marked stimulation of growth of ganglion cells *in vitro* and *in vivo* is produced by microgram quantities of the material. When antiserum to the factor is administered to newborn mice, it destroys the sympathetic ganglia. The molecule is comprised of three subunits, only one of which is active. The active unit is essential for growth and differentiation of neuroblasts in tissue culture, but its role in the adult organism is not known (Levi-Montalcini, 1975).

Epidermal growth factor

In studies with nerve growth factor, Taylor, Cohen, and Mitchell (1970) observed premature opening of the eyes and eruption of incisors in mouse embryos. These effects were due to the presence of epidermal growth factor in the preparation. The epidermal growth factor has proved to be a polypeptide of from 6,000 to 7,000 molecular weight, which is sometimes bound to a carrier protein of approximately 30,000 molecular weight (Taylor, Cohen, and Mitchell, 1970). Like nerve growth factor, epidermal growth factor is active in microgram quantities *in vivo*, and in tissue culture it stimulates uptake of tritium-labeled thymidine and migration of epidermal cells. Epidermal growth factor binding protein is an arginine esterase, and since the epidermal growth factor has a carboxylterminal arginine residue, it may be generated from a precursor protein by proteolytic action.

Nerve growth factor and epidermal growth factor may be closely related biochemically, and they may be under the control of the same genetic locus. It is interesting that the application of epidermal growth factor appears to increase the carcinogenicity of certain molecules. It is conceivable that by stimulation of mitotic activity either more target cells are made available for initiation or initiated cells may be promoted.

Embryo extracts

Experimentalists have traditionally used extracts of embryonic chick tissues as growth stimulators in the establishment of tissue cultures. Coon and Cahn (1966) used exclusion chromatography to fractionate extracts of 9-day-old chick embryos and tested their biological effects on the growth and differentiation of retinal pigment cells and on cartilage cells in culture. A high molecular weight fraction of the extracts inhibited the differentiation *in vitro* of pigment cells and cartilage-forming cells, but the low molecular weight component allowed expression of the differentiated function. They also found that the state of differentiation of these cells could be altered by the presence or absence of extract.

Serum factors

Serum-free, chemically defined media have been adapted for certain cells in culture, but they are not generally capable of sustaining the growth of cells plated at low densities. Serum contains a low molecular weight protein that has been purified and it has been shown to control the final cell density of 3T3 mouse cells. There are also two heat-labile factors that are involved in cell division and growth (Holley, 1974). There is a serum factor required for survival of 3T3 cells, a factor that influences migration of the cells, and a factor that is involved in transport of phosphate ions. Temin, Smith, and Dular (1974) have described a multiplication-stimulating activity (MSA) that has a molecular weight of 10,000. This material is insulin-like, but the activity is not reduced by insulin antibodies. MSA resembles somatomedin and its activity requires the presence of both amino acids and glucose. Cells transformed by Rous sarcoma virus (RSV) require less MSA for multiplication than normal cells, but they die without MSA. In general, transformed cells have lower requirements for serum than normal cells when grown in culture.

It has been difficult to assess the effects of various serum and embryonic growth factors in tissue culture because of the complexity of the nutritional requirements of cells and the presence of unknown quantities of these factors in cells. Attempts to determine the relative importance of each of the growth factors is hampered by difficulties in preparing media deficient in one factor but containing optimal concentrations of the other factors. Growing cells with either feeder layers or cells in conditioned media frequently improves the cloning efficiency of cells in culture. This effect may be related to the utilization of certain nutrients (*e.g.*, serine) that can leak from cells, but there are other possible explanations.

Low molecular weight compounds

Cells in culture can be arrested in G_1 by limiting certain amino acids. Phosphate limitation also arrests growth; this can be reversed with phosphates, but only in the presence of a serum factor that stimulates phosphate uptake. The requirements of potassium, calcium, magnesium, and zinc ions in cell replication have been studied. It is of interest that calcium ion deprivation has a much greater effect on the growth of chicken cells than on RSV-sarcoma-virus-transformed chicken cells.

The pH of the medium markedly alters the growth of cells in culture (MacKenzie, MacKenzie, and Beck, 1961; Eagle, 1974). Most virus-transformed cells have a more acid pH optimum. The pH of medium for normal cells in culture can be so optimized that there is a marked increase in the final saturation density comparable to that observed for transformed cells.

The role of hormones in growth control is dependent upon both the hormone and target cell. Cortisol stimulates DNA synthesis and cell division in density-inhibited fibroblasts in culture, but transformed fibroblasts are unresponsive to cortisol. In general, the more anaplastic tumors are not responsive to normal hormonal control. For example, adrenal cortical carcinomas are frequently insensitive to ACTH with regards to secretion of adrenal hormones.

Although certain hormones stimulate the proliferation of some cells, in other situations they have a lethal effect. Dexamethasone, for example, has a lethal effect on most lymphoma cells. Sibley *et al.* (1974) have been able to isolate lymphoma cells that are resistant to the killing action of such steroids. Because of a defect in their steroid receptors, these cells are resistant to the effects of steroids.

The molecular action of hormones on cells in culture is being studied with respect to receptor molecules, cAMP levels, glucose transport, and transcriptional and translational control mechanisms. The hormonal stimulation of growth in some tissues is thought to be mediated by cyclic AMP, the *second messenger*. When serum factors attach to the cell membrane, they lower adenyl cyclase activity and they secondarily decrease cAMP levels. It is speculated that the level of cAMP regulates cell growth and morphology by controlling the transport of nutrients such as glucose into the cell. Thus, cyclic AMP seems to act as an inhibitor of growth. In transformed cells the property of unrestrained growth may result from membrane alterations that allow increased transport of nutrients such as glucose, amino acids, and phosphate and that allow depression of cAMP. The levels of cellular cAMP may also be important in differentiation. Prasad (1973) has shown that mouse neuroblastoma cells slow their growth rate and undergo morphological differentiation when treated with agents such as prostaglandin E and phosphodiesterase inhibitors. These increase the intracellular cAMP.

The role of cyclic GMP in cell proliferation is not well understood, but it appears that when the levels of cAMP decrease, the levels of cyclic GMP increase. When lymphocytes are stimulated to divide by plant lectins such as phytohemagglutinin, the cAMP levels decrease and the cyclic GMP levels increase. Cyclic GMP at extremely low concentrations (10^{-11} M) stimulates RNA synthesis in lymphocytes. The promoting substance, phorbol ester, which is capable of inducing proliferation in fibroblasts, causes a dramatic increase in cyclic GMP within 60 seconds after application. Cell proliferation may be related to a balance of both cyclic GMP and cyclic AMP levels (Goldberg *et al.*, 1974).

Tumors in developing systems

The concept of individuation fields proposes that regions of embryos exert controls upon the cells within the region (Needham, 1936). It has been proposed that the development of neoplasms might represent an escape on the part of certain cells from these controls. If so, individuation fields must persist into adulthood. Amputation of the limb of certain amphibians results in the formation of an undifferentiated blastema that then reorganizes and differentiates into a new limb. If carcinomas were an escape from the effect of a field, then an appropriate tumor placed in a regenerating limb might be influenced by the regenerating field. The neoplastic cells might even be incorporated into the field as normal components.

The effect of fields on cancer cells as postulated by Needham (1936) stimulated experiments by Seilern-Aspang and Kratochwil (1962). As discussed previously, they induced tumors in newts with chemical carcinogens and found that many of the tumors spontaneously regressed. This, of course, is characteristic of chemically induced tumors in a variety of systems and cannot be attributed to field effects. In addition, some of the cells of their malignant tumors differentiated into pigment cells, mucous glands, and cutaneous epithelium. This has been interpreted by some as control of malignant growth by the host, but since it is now known that differentiation occurs spontaneously in tumors, the interpretation would probably not be accepted by most oncologists.

Ruben (1956) found no evidence of a field effect when he grafted lymphoma tissue onto regenerating limbs of amphibians; the growth and organization of the tumors were unchanged. This might be expected because the regenerating limb is histogenetically different from lymphoma tissue. A more appropriate test would have been the implantation of fibrosarcoma cells in the blastema.

More recently, de Lustig and Matos (1972) implanted crystals of chemical carcinogen in the tails of tadpoles and, upon the appearance of lymphomatous growths, amputated the tails distal to the point of tumor growth. As regeneration progressed, the lymphoma nodules were encapsulated, underwent degenerative changes, and were apparently resorbed. In some situations, the undifferentiated tumor cells piled up and formed epithelial aggregates with gland-like patterns. There are several possible explanations for the gland-like epithelial cells in these tumors. It is conceivable that they could have evolved directly from the action of the carcinogen on epithelium and their presence prior to the regenerative phenomena could have been masked by the presence of the lymphoma. It is unlikely that lymphoid cells could differentiate into gland-like structures. In similar experiments de Lustig *et al.* (1968) cultivated mammary adenocarcinoma cells on either the primitive streak of 20-hour-old embryos or the notochords of $3\frac{1}{2}$-day-old chick embryos and observed acinar formations containing secretions. The explanation for this effect and its specificity are not known. Of great interest is a conclusion that can be made from these experiments: The embryonic field can apparently overcome the neoplastic field.

Brinster (1975) first produced an allophenic mouse by incorporation of a single embryonal carcinoma cell into the blastocyst of a foreign strain and allowed development of the grafted embryo to proceed to term in a pseudopregnant animal. These experiments were elegantly extended by Mintz, Illmensee, and Gearhardt (1975), who demonstrated that the incorporated embryonal carcinoma cells differentiated and took part in normal development. The resultant tissues appeared normal and showed no neoplastic features. Papaionnou, McBurney, and Gardner (1975) grafted larger numbers of embryonal carcinoma cells into blastocysts and although allophenic mice were produced, there was also residual tumor in the offspring. Whatever the stimulus in the blastocyst, it was able to direct the differentiation and incorporation of offspring of embryonal carcinoma cells into the normal organism. If too many embryonal carcinoma cells were present, the effect was insufficient and some of the cancer cells persisted as cancer cells. This again points out the importance of critical numbers of cells creating their own appropriate environment for expression of a particular phenotype. Apparently when a few cancer cells were used, the microenvironment resulted in signals that influenced the genome to behave in a normal manner. Whether or not this represents the effect of an individuation field is moot.

These examples point out the kinds of experiments that might be done to test the effects of appropriate environments on controlling the malignant phenotype. Since embryonal carcinoma cells are the equivalent of a $4\frac{1}{2}$-day embryonic mouse epithelium, implantation of these cells into blastocysts provide an appropriate environment. Attempting to study the effects of regenerating limb blastema on a grafted lymphosarcoma seems inappropriate because normal lymphoreticular tissue does not participate in regeneration of a limb. Had fibrosarcoma cells been incorporated in the blastema, an allophenic limb might have resulted, and it would be interesting to see whether such cells might have given rise to non-neoplastic fibroblasts. Similarly, it would be inappropriate to place embryonal carcinoma cells on either the notochord or on the head blastema, but it would be appropriate to place rhabdomyosarcoma cells in a somite or a chondrosarcoma cell in a chondrogenic part of a somite. Molecular explanations of what happens to the embryonal carcinoma cell when placed in a blastocyst will probably represent the prototype for a new therapy of cancer.

Whisson (1972) considers carcinogenesis to be an extension of the normal processes of differentiation and offers the plasmagene theory of inheritance as its explanation. He believes that deletion of cytoplasmic factors (plasmagenes) resulting from unequal cell division is the mechanism of differentiation. Tumors would arise by a similar mechanism. If tumor cells were placed in embryonic tissues that had a full set of plasmagenes, these plasmagenes might be able to influence the differentiation of the tumor cells and cause them to become benign. At any rate, Whisson quotes evidence to the effect that "tumor cells do not grow well when injected into embryos." He implanted Yoshida ascites tumor cells into rat embryos between the eighth and eighteenth day of gestation and found that there was a marked inhibitory effect on growth of tumor cells. He believed that there was organization of tumor cells where the tumor cells contacted normal mesenchyme. On the other hand, when non-neoplastic embryonic cells were injected into growing tumors, the embryonic cells differentiated and there was apparently little effect upon the tumors. It is difficult to interpret these experiments, but it is worth noting that Whisson raises the question of whether or not tumor cells are an abnormal type of, or an extreme extension of, the normal process of development. More refined experiments of this type are warranted.

Of more immediate concern is the recognition of the fact that the local environment of cells is important in carcinogenesis. Dawe (1972) has shown that appropriate epithelial-mesenchymal interactions are required for differentiation of the submandibular salivary gland. If glands were exposed to polyoma virus at a relatively advanced stage of development, they invariably gave rise to polyoma-induced carcinomas. If, however, glands were removed earlier in embryogenesis and exposed to polyoma virus, few tumors developed. Those removed at the earliest developmental stage never resulted in tumors. Dawe concluded that the neoplastic potential of salivary gland epithelium depended upon the state of organogenesis of the gland. In addition, when epithelial and mesenchymal components of salivary gland rudiments were separated from each other and each was infected with polyoma virus, no transformation occurred. When the epithelium and appropriate mesenchyme were recombined and exposed, transformation took place as in the normal gland.

These data lead to the conclusion that both normal and neoplastic differentiation are dependent upon mesenchymal–epithelial interactions. Dawe postulates that polyoma-induced salivary tumors originate after epithelium acquires the ability to produce growth-supporting factors that are normally provided and regulated by mesenchyme. It would also follow that the mesenchymal effect, once developed in normal tissue, persists and allows for development of tumors in adult animals.

In conclusion, the stability and heritability of malignant change can be explained either by somatic mutation or by differentiation. The concept of carcinogenesis as the equivalent of a differentiation places neoplasms in the mainstream of biology, accounts for all of the characteristics of neoplasms, and corresponds with what is known about cytoplasmic control of the genome. We favor this idea over somatic mutation.

It is apparent from the foregoing account that intensive study of the process of differentiation in mammals is urgently required. This information should elucidate the extracellular factors that allow for differentiations to occur and the cytoplasmic factors that allow for control of gene expression. This information should lead to a new therapeutic approach to cancer.

NINE

Immunity
and Neoplasms

Introduction

Immunity, or an immune response, implies an actively acquired defense on the part of the host, mediated by cells (the immunocytes include lymphocytes, plasma cells, and macrophages) and/or circulating antibodies (the various classes of immunoglobulins) in response to antigens recognized as foreign by the individual. Antigens include viruses, bacteria, tissues, and many kinds of proteins, carbohydrates, and lipids (Sell, 1975).

The humoral or circulating antibody response is typical of most common bacterial infections. Bacterial antigens cause bone marrow-derived lymphocytes to proliferate and differentiate into plasma cells that then synthesize various classes of immunoglobulins specific for the bacterial antigens. The precise mechanism of this reaction is not known. What is known is that specific antibodies combine with their antigens and render the bacterium susceptible to phagocytosis and intracellular digestion by leukocytes. Concomitantly, toxins produced by the bacterium also serve an antigen, elicit an antibody response, and combine with these antibodies (antitoxins) to form insoluble precipitates that are removed from the circulation by the reticuloendothelial system.

Rejection of tissue transplants is an example of cell-mediated immunity. When skin is transplanted from a donor mouse to an unrelated recipient, it appears healthy for from 10 to 12 days and then it becomes discolored and dies. When a second graft from the same donor is applied, it is rejected in 3 or 4 days, indicating that prior sensitization enhances the immune response. This type of immunity is mediated by small lymphocytes that are thymus-derived rather than bone marrow-derived. The precise manner in which these cells react with the graft *in vivo* to cause its destruction is not known, but from *in vitro* studies it would appear that sensitized small lymphocytes interact directly with target cells in the graft. A humoral response to antigens found in the cell membranes of the donor tissue usually accompanies transplant rejection. Like other antibodies, it reacts specifically with its antigen on cell membranes, and in the presence of complement it will cause destruction of target cells in tissue culture. Interestingly, if this cyto-toxic antibody directed against skin cells is injected into an animal carrying a transplant of those skin cells, it does not destroy the graft.

Immune surveillance

If tumor cells have surface antigens recognized as foreign by the host, then why shouldn't all tumors be rejected? Thomas (1959), in philosophizing about cellular immunity, concluded that it had not evolved over the ages in order to confound transplant surgeons or to combat a few types of bacteria such as tuberculosis or leprosy. He postulated that cell-mediated immunity was useful in ridding the body of aberrant cells. This notion, extensively developed by Burnet (1971), has been named *immune surveillance*. According to this concept, foci of neoplastic cells periodically arise, are recognized as antigenically *foreign* by the host, and rejected—much as a homograft is rejected. If this system of immunologic surveillance breaks down, clinical cancer evolves. This concept is supported by the fact that the experimental production of many tumors is facilitated in immunologically immature animals or in animals made immunodeficient by various procedures (*e.g.*, neonatal thymectomy, radiation, or the use of immunosuppressive drugs). In human beings there is an increased incidence of cancer in persons with primary immunodeficiencies (Kersey, Spector, and Good, 1973) and in homograft recipients who have been made immunodeficient (Penn and Starzl, 1972). Prehn (1974), however, although not denying a role for immune mechanisms in cancer biology, notes that "the evidence in favor of immunologic surveillance is weighty but circumstantial." He notes, for example, that although the congenitally athymic *nude* mouse is incapable of rejecting homografts, it does not appear to show any increased susceptibility to experimentally induced cancer—except that of the lymphoreticular system.

Thus, the essence of tumor immunology is the concept of immune surveillance and the end point is to determine if and how the reticu-loendothelial system can rid the body of incipient tumors.

If immune surveillance really works and destroys small numbers of aberrant cells as they develop, we would never know that the reaction had successfully occurred unless some kind of marker were left behind. Because we are dealing with immunologic reactions, the best markers would be either antibody or sensitized cells directed against tumor-specific transplantation antigens. Since it has been shown that chemical carcinogens induce different antigens in each tumor (Prehn and Main, 1957), and sometimes multiple antigens in the same tumor, there is no way to test whether sensitized cells or antibodies induced by the tumor antigen might be present after destruction of a few chemically initiated stem cells, because the only antigens capable of reacting with the cells or antibody would have been destroyed. Interestingly enough, the tumor-specific transplantation antigens (TSTA) induced by oncogenic viruses are specific for the virus, irrespective of the kind of tumor or the species in which it is induced. Thus, if a few incipient tumor cells were induced by polyoma virus, for example, and then destroyed as a result of immune surveillance, one should expect some antibody to polyoma-tumor-specific transplantation antigens to persist and be easily detected.

If polyoma virus is injected into an adult animal with no evidence of circulating antibodies to polyoma-tumor-specific transplantation antigens, a tumor does not develop. In contrast, if polyoma virus is injected into a newborn animal, tumors do develop. The different outcomes can be ascribed to immune surveillance. We know that there was immune surveillance in the adult because animals previously nega-tive for circulating antibody to polyoma TSTA became positive after exposure to the virus.

If an established tumor induced by polyoma virus in a particular strain is passaged into other members of the same inbred strain, the tumors do not grow (Lehman, personal communication, 1977). If an animal in a particular inbred strain carrying a polyoma-induced tumor is given lethal irradiation, and its immunopoietic system is reconsti-tuted by transplantation of sensitized immunocytes, the tumor is re-jected. The same situation holds true for tumors induced by SV_{40}.

Several conclusions can be drawn from these observations. In the first place, it may be concluded that the concept of immune surveillance has been established, the opinion of Schwartz (1975) to the contrary. In the second place, the TSTAs induced by polyoma virus must be excellent ones because immune surveillance works against them. Even large established polyoma tumors can be rejected by suitably sensitized cells directed against these antigens. The TSTA of polyoma and SV_{40}

virus may be so strong that the minor antigenic changes induced simultaneously in cell membranes may be masked.

Before we discuss these data further it would be wise to consider the development of chemically induced tumors in relation to tumor immunology. Foley (1953) demonstrated that a methylcholanthrene-induced sarcoma contained TSTA. Animals bearing transplants of this tumor were operated upon and vessels supplying the tumor ligated. After the tumors regressed, the animals were reinoculated with sarcoma cells from the original tumor and the incidence of "takes" was found to be lower than that in animals that had not been previously exposed to the tumor. Failure to achieve immunization by using normal tissues indicated that the mechanism was mediated by antigens peculiar to the sarcoma. These observations were extended by Prehn and Main (1957) who demonstrated that after surgical removal of a tumor, an animal was refractory to the inoculation of cells from the same tumor. Even large numbers of tumor cells were rejected in this system. Prehn and Main also discovered that, unlike the situation with oncogenic viruses, the chemical carcinogens induced tumors with individually tumor-specific transplantation antigens. Thus, an animal that had rejected one methylcholanthrene-induced sarcoma would not be refractory to a different sarcoma produced in the same inbred strain by the same carcinogen. These data were confirmed by Bosombrio (1970) who examined the antigenicity of methylcholanthrene-induced sarcomas in 25 mice of a single strain. Each of the sarcomas was antigenically distinct not only from the host's non-neoplastic tissues but also from each other.

Chemical carcinogens are not informational molecules in the sense that oncogenic viruses contain genetic information. Thus, it can be assumed that the chemical carcinogen stimulates the host cell to produce these transplantation antigens. This means that they are encoded for in the normal genome and one wonders what their function might have been during development or in ontogeny. Several interesting problems are posed by these experiments. Application of chemical carcinogens results in initiation of latent neoplastic stem cells, which one would expect should be destroyed by the mechanisms of immune surveillance. The tumor cells that develop contain tumor-specific transplantation antigens; yet they grow in the autochthonous host and produce tumors. If the tumor is surgically removed, the host will reject subsequent inocula of the identical tumor. Thus, explanations must be found for the establishment of a tumor that can elicit cytotoxic reactions and yet survive in the presence of those reactions.

Chemical carcinogens are highly cytotoxic and it has been shown that when many of them are parenterally administered, they nonspecifically depress the host's immune response. Thus, it could be argued that immune surveillance is depressed by the carcinogenic agent. It is unlikely that the immune system would stay depressed during the entire latent period which may be half of the life span of the animal. In addition, local application of carcinogen as in the case of painting chemical carcinogen on the skin would be unlikely to cause chemical depression of the immune response. The explanation for the development of a tumor, *i.e.*, the events following initiation, probably resides in the quantity and quality of the tumor-specific transplantation antigens involved. Baldwin (1973, 1975) has shown that not all tumors carry tumor-specific transplantation antigens. In the case of the polyoma and SV_{40}-induced tumors, tumor-specific transplantation antigens are good antigens and the tumors are rejected unless the animal develops initiated cells during a period when it is immunologically incompetent and develops tolerance to the foreign antigens. Thus, tumors might develop if TSTA was of poor quality.

Immune enchantment

Kaliss (1969) has described a phenomenon now known as *immune enhancement*. In this situation, animals that would normally reject a transplant of a particular tumor have enhanced growth of the tumor when treated with cytotoxic antiserum developed against that tumor. Not only did the tumors grow, but they also grew more rapidly than one would have anticipated. Such enhancement of growth has been attributed to blocking antibodies (Hellstrom and Hellstrom, 1970). The mechanism is not known, but it appears to be related to an interaction between circulating antibody and antigen on the surface of the cell, thereby interfering with cell-mediated rejection. Since tumors are known to shed membrane associated antigens into the circulation, and, in particular, glycoprotein antigens, one wonders whether or not enhancement might also be due to circulating antigen that might tie up the immune system.

Tumor-specific and transplantation antigens

There is experimental support for the idea that an established tumor represents an antigenic challenge too large for the animal to handle. Since transplanted livers can be rejected, this explanation may seem unlikely. However, organ rejection seems to be the result of immune rejection of the blood vessels of the donor tissue resulting in necrosis of the organ. Thus, the situations are not parallel.

Occasionally, renal homografts have been done in human beings in which the donor kidney inadvertently contained tumor cells. In the immunologically suppressed recipient, the kidney thrived and the tumor cells grew and ultimately metastasized. With the restoration of immunocompetence, by cessation of immunosuppressive therapy, both kidney and tumor were rejected. Since in this situation the donor and recipient are not isogeneic, it is unknown whether tumor rejection was due to TSTA or simply due to the same transplantation antigens responsible for rejection of the kidney. Another clinical situation in which transplantation rejection may play a role is choriocarcinoma (Bagshawe, 1973). Postgestational choriocarcinoma of the uterus is derived from fetal trophoblast. This is a highly malignant tumor that, until recently, has been almost uniformly fatal despite surgery and radiotherapy. The application of chemotherapy has markedly improved the prognosis of these patients, such that approximately 70% of cases are now curable. Testicular choriocarcinoma, however, is resistant to all forms of therapy. Uterine choriocarcinoma is derived from fetal troblast, tissue that contains both maternal and paternal antigens and is, therefore, equivalent to a homograft. Testicular choriocarcinoma, however, is derived from the host's own tissues. When postgestational or testicular choriocarcinomas were heterografted to immunosuppressed hamsters, the tumors responded equally well to chemotherapy. It is suggested that the differences in chemotherapeutic response seen in the original hosts might reflect their different TSTAs.

In human beings, immunologic specificity may be determined more by the type of tumor than anything else. This is different from the etiologically related antigenic specificity of viral-induced experimental tumors and the individual tumor specificity of experimental chemically induced tumors. Similar host immunologic reactions and tumor-type antigenic specificity have been reported among human melanomas (Morton, Eilber, and Malmgren, 1970), lymphomas (Golub *et al.*, 1972), leukemias (Herberman *et al.*, 1972), and sarcomas (Vanky, Stjernsword, and Milsonne, 1971). These observations are being applied experimentally to the therapy of these tumors in humans (see Chapter Eleven).

Finally, although tumor antigens and host responses may play a role in the clinical course and therapy of *established* tumors, there is increasing evidence that caution must be used in considering such antigens as truly tumor specific. Antigens that had been called TSTAs (*i.e.*, unique to tumor cells and absent from non-neoplastic cells) have been demonstrated in embryonic and other rapidly growing, non-neoplastic tissues. For example, carcinoembryonic antigen, originally demonstrated in human colonic carcinoma by Gold and Freedman (1975), has been found in embryonic and fetal gut, normal childhood and adult colon, and other endodermally derived, non-neoplastic and neoplastic tissues (Zamcheck, 1975). Experimentally, Coggin *et al.* (1971) found that prior immunization with fetal tissues enhanced rejection of subsequently transplanted SV_{40} and adenovirus-induced tumors in hamsters. They suggested that tumor antigens might really be *phase-specific* antigens present at sometime during normal development (Coggin and Anderson, 1974). Many tumor immunologists are becoming increasingly reluctant to use the term tumor-specific antigen, preferring instead the less committal term *tumor associated antigen.*

It is likely that many so-called tumor-specific antigens will turn out to be phenotypic expressions of normal stem cells rather than cancer-specific phenomena. Their presence may simply reflect our ability to detect them. Thus, relatively insensitive methods might detect them only under conditions in which stem cells are abundant. In the case of carcinoembryonic antigen, they are detectable in developing, regenerating, or neoplastic colon. With more sensitive methods, *i.e.*, in situations in which stem cells are less numerous, they are found in normal colon as well. Thus, perhaps these antigens should be called *stem cell specific* rather than *tumor specific*. As in the case of biochemical, morphologic, and other parameters, any search for biologically unique "handles" should compare neoplastic with normal stem cells. The possibility of antigenic similarity between normal and neoplastic stem cells suggests that caution should be observed in therapy directed against tumor-specific antigens.

TEN

Metastasis

Introduction

Metastasis is the process of dissemination of malignant cells from a primary tumor with growth of secondary, or metastatic, tumors in other tissues and organs. Invasion and metastasis are the most feared consequence of tumors, and their presence is viewed as unequivocal evidence that the tumor is malignant. People with metastasis are usually doomed, and for the most part, treatment is palliative because the metastases are multiple, often widespread throughout the body, and therefore not amenable to x-ray or surgical therapy.

Control of the process of metastasis could reduce the hazards of malignant tumors to about that of benign tumors (positional–functional effects) that are usually handled easily. An enormous amount of clinical observation has been accumulated about metastasis, but it has yielded more questions than solutions. Basic investigation has been of less help than anticipated because there is no good experimental model of metastasis that answers the clinical questions.

Clinical features

So many patients have now been studied that certain generalizations can be made. Metastasis is not a property of the few cells initiated by chemical carcinogen; instead it is an attribute that is expressed by cells after the development of the neoplastic mass. In general, the smaller the mass, the less likely it is that metastasis will

159

have occurred, but there are many exceptions to this rule. For example, minute teratocarcinomas of the testes may be accompanied by widespread metastases, while some malignant tumors of the kidney or breast attain huge size without metastasis. It is more generally true, however, that the larger the tumor the greater the likelihood of metastasis (Willis, 1973).

Statistics are available on the frequency of metastatic sites for every kind of tumor. For example, adenocarcinoma of the breast metastasizes most frequently to regional lymph nodes, then to lungs, liver, bones, adrenals, and ovaries. Adenocarcinomas of the stomach metastasize to regional lymph nodes, then to liver and lungs. It is not uncommon for malignant cells of adenocarcinomas of the stomach to invade the stomach wall and seed the peritoneal cavity with the formation of an ascites (p. 169). Adenocarcinomas of the prostate have a predilection for metastasis to bone, particularly those of the spine. It turns out that venules draining the vertebral column anastomose with those draining the prostate. Normally blood flows from the spine toward the heart, but when intraabdominal pressure is raised by lifting or straining, the flow is reversed and malignant cells may be carried from the prostate into the spine. Lung cancers metastasize to the brain frequently enough so that patients with signs of a brain tumor are evaluated to rule out a metastasis from the lung. This propensity of tumors for metastasizing to certain organs has long been recognized. Paget likened the metastasizing tumor cells to seeds and the recipient organs to soil and suggested that certain kinds of seeds require a particular soil for successful growth (Willis, 1973).

Other tumors, such as melanoma and lung cancer, are unpredictable in their pattern of metastasis. Whereas most metastases are multiple and often widely disseminated, some tumors may produce only a single focus. Adenocarcinomas of the kidney may metastasize as a single "cannonball" in the lung. Malignant tumors of the brain rarely metastasize outside of the central nervous system, yet their capacity for growth in other locations is illustrated by implantation of these cells, sometimes with secondary metastasis, as a result of accidental seeding at the time of surgery. Unfortunately, it is not uncommon for a primary tumor to be first recognized as a result of disease produced by its metastases.

It is apparent that tumors can metastasize in three ways: through lymphatics, through blood vessels, or by implantation in the body cavities. To gain access to any of these conduits, it is necessary for malignant cells to break away from the primary mass and invade a vessel or body cavity. Then they are transported to the site of implantation where they may form a new tumor. The established metastasis usually resembles the primary tumor histologically and may grow to a size greater than that of the primary tumor. Its cells may invade and metastasize again, and the patient dies either of destruction of a vital structure or by cachexia and infection.

Invasion-migration

(Pg 168)

No suitable explanation for cells breaking away from a parent mass, invading, and metastasizing has been found. As pointed out in Chapter Five, these properties are not unique to malignant cells and they correspond directly to migration in the embryo (Trinkaus, in preparation). In the embryo the widespread migration of neuroblasts, melanoblasts, and primordial germ cells is well known. Lymphocytes produced in the bone marrow metastasize via the bloodstream to their ultimate locations in lymphoid tissues throughout the body where they serve their immunologic roles. As in the case of other properties of cancer cells, it appears that metastasis is not a matter of uniqueness, but rather of altered control of a normal process. Migration in the embryo is under stringent, although as yet unknown, controls, but in tumors the process of widespread dissemination, plus the continuing capacity for proliferation of tumor cells, although not uncontrolled, ultimately results in death of the host (Wallace, 1961).

Before cells can invade and metastasize they must be able to separate from the primary mass. Of interest in this regard was Coman's demonstration that tumor cells were less adhesive to each other than were their normal counterparts. Coman (1944) measured the mechanical force necessary to pull adjacent cells of a squamous cell carcinoma apart; he found that it was less than that required to pull normal cells apart. Coman deduced that an alteration in cell membranes was responsible for the decreased adhesion between tumor cells, and he and his associates demonstrated that membranes of tumor cells contain less Ca^{++} than those of most normal cells (1953). Specifically, the plasma membrane of human intestinal cancer was found to contain 45% less Ca^{++} than did those of normal mucosa. This difference in Ca^{++} content of malignant versus normal cells has never been explained, but cell biologists have taken advantage of this observation by using chelating agents to dissociate cells from monolayer cultures (Lilien and Moscona, 1967). There is less adhesion between transformed cells *in vitro* and their substrate than between normal cells *in vitro* and their substrate (Dorsy and Roth, 1973). But there appears to be little relationship between cell adhesion in a tumor and the ability of the tumor to metastasize. The precise role of cell adhesion in metastasis has yet to be determined, but it pinpoints cell membranes and intercellular substances as potentially important factors in the process.

It may seem odd to refer to the social relationships of cells, but there is now no question that such a consideration is appropriate to the understanding of migration invasion and metastasis. Wilson, in 1907, showed that mechanically dissociated sponge cells reaggregate, and Townes and Holtfreter (1955) demonstrated that mixtures of dissociated heart, muscle, and liver cells, when left to their own devices *in vitro*, had a tendency to sort themselves out in the process of aggregation. Heart cells aggregated with heart cells, liver cells with liver cells, and muscle cells with muscle cells. More recently, Moscona (1961, 1962, 1963) described many of the characteristics of this sorting out and has related the findings to histogenesis.

Moscona dissociated embryonic tissues with EDTA and also observed that cells from like tissues had a tendency to aggregate from suspensions of single cells. This aggregation occurred even when tissues from different species were used. Tissue affinities were found to be much stronger than species affinities (Moscona, 1957). In addition, certain cells aggregated in specific patterns. For example, mixtures of dissociated cartilage and muscle cells reaggregated to form a central nodule of cartilage surrounded by a layer of muscle. When tumor cells were added to the aggregating masses, they did not show any particular affinities. Possibly tumor cells lack the recognition factors possessed by non-neoplastic cells. Weiss (1950) noted this phenomenon in the formation of specific neural connections. Konigsberg's demonstration (1965) of fusion of myoblasts to form myotubes indicates the specificity of cell contacts in tissue genesis.

The mechanisms of aggregation have been studied in detail. Lilien (1968) and Moscona (1962, 1963) believe that an extracellular material, antigenic and apparently membrane-bound, plays an important role. Moscona has shown that aggregation is stopped by agents that block protein synthesis. Actinomycin, which blocks transcription, and puromycin, which blocks protein synthesis at the translational level, are both effective in preventing reaggregation. These observations are compatible with the idea that dissociation of cells removes essential molecules and that these must be replaced before aggregation can take place. It has also been suggested that much of the extracellular material is denatured DNA, but, interestingly, cells still reaggregate after treatment with DNAse. Adhesive factors have been isolated by Orr and Roseman (1969) and McClary and Moscona (1974), and it might appear that aggregation is an active process. Umbreit and Roseman (1975) have dissected the problem into two parts. Upon initial contact, liver cells formed a loose association that was insensitive to cyanide and these cells were easily dissociated by shearing. After a second, more stable kind of bond, the cells were shear resistant. This step was inhibited by cyanide, suggesting that it was energy dependent or that a membrane junction had to be formed. Since the mechanisms by which cells are held together are unknown, the mechanisms by which they become dissociated and invasive cannot, as yet, be deduced. This is an area of great scientific importance.

The characteristics of growing populations of cells have been described *in vitro*. When a sufficient number of embryonic fibroblasts are inoculated into a tissue culture plate, they cover the surface as a uniform monolayer adherent to the substrate (*anchorage dependency*). When Abercrombie and Ambrose (1962) studied the social behavior of these cells *in vitro*, they observed that there was an inhibition of further movement (*contact inhibition*) when normal fibroblasts made contact. Thus, normal fibroblasts respect the presence of other cells and do not encroach upon their territory. Growth is also controlled, such that when the surface is covered by a monolayer (*saturation density*), the cells stop synthesis of DNA. Thus, normal cells are able to regulate their growth and social relationships.

Neoplastic cells go beyond the monolayer state (lack of contact inhibition) (Leighton, 1957), slough easily from the plates and grow in semisolid media (lack of anchorage dependency), and continue to synthesize DNA after the saturation density of normal cells has been exceeded. Understanding of these changes from normal behavior of cells will undoubtedly shed light on the understanding of invasion and metastasis (see Transformation, p. 98).

Transformed cells are more agglutinable, with lower concentrations of plant lectins, than are normal ones. Trypsinized normal cells become more agglutinable, suggesting that receptor sites for lectins might be masked on normal cell surfaces. It is now known that the number of lectin binding sites on normal and transformed cells is similar (DePetris, Raff, and Mallucci, 1973) and differences in agglutinability may have more to do with differences in mobility of receptor sites (Yahara and Edleman, 1975) than their number (p. 63).

One of the most important means of studying invasion has been Leighton's three-dimensional tissue culture system (1957, 1967) in which it is possible to observe the interactions of tumor and other cells *in vitro*. Cells to be tested were introduced into a thin slab of photographic sponge and supported within the interstices of the sponge by a plasma clot. Many clinical problems were studied by using this system, but of particular importance were observations of the social relationships of tumor cells and invasion. Cells from some tumors migrated over or around normal cells while other tumor cells formed aggregates that behaved independently. The aggregate could migrate or divide; in other words, once a critical number of cells was reached, the aggregate of cells functioned as a unit. The behavior of the aggregate depended upon the particular strain of tumor from which it was derived. Possibly, Leighton was measuring the organizational capacity of cancer cells, some requiring a critical number of cells to operate optimally and others manifesting the malignant phenotype as single cells. At any rate, other tumor cells appeared to operate optimally as groups, invading and probably metastasizing as aggregates. The aggregate is thus a unit of efficiency, and it may represent the threshold number of cells sufficient to create a socially acceptable milieu for expression of the phenotype of its component cells. Partial explanation for the phenomenon observed by Leighton may be found in the experiments of Grobstein and Zwilling (1953) that indicated that a critical mass of cells was required for differentiation (p. 18). Unfortunately, we still don't know how the cells create the optimal environment.

From an ultrastructural standpoint, epithelial cells are often held together by desmosomes and other types of junctions. The importance of these structures in cellular aggregation or in tissue genesis is not known. The nexus is a type of junction across which electrical currents flow and which can be broken by various kinds of stimulation. Jamarosmanovic and Lowenstein (1968) have described intercellular communication between like cells that allow the passage of molecules. It is conceivable that the passage of informational molecules between cells plays a part in the development, maintenance, and functions of tissues. Whether or not desmosomes and other junctions play a role in metastasis is not known. Some have reported normal numbers of intercellular communications in tumors, but others have reported diminished numbers (see Electron microscopy).

Granted that a critical number of cells is necessary to create an optimal microenvironment, it still does not explain why tumor cells should dissociate and migrate away from the parent mass. It has been suggested that these cells are seeking more optimal metabolic environments. Others believe that neoplasms grow under pressure and that this pressure might squeeze neoplastic cells along tissue planes (lines of least resistance); again, this is not a satisfactory explanation for the phenomenon of invasion. Tumors have been examined for the presence of enzymes capable of lysing intercellular substances. Some tumors that metastasize widely contain hyaluronidase; others that metastasize just as widely do not.

Germ cells of embryonic birds invade the vasculature and are widely disseminated, but they colonize only the gonads. Thus, there appear to be controls, or at least signals, that specify where these cells should colonize. Weiss and Andres (1952) injected pigmented feather follicle cells intravenously into albino chick embryos and obtained birds with partially pigmented feathers; all other tissues remained unchanged. This suggests that circulating feather follicle cells may have responded to specific signals in settling out and colonizing in feather follicles. If these signals occur, they must be diffusible, at least across a basement membrane, and have a short range of reactivity in the vasculature. Of great importance is the fate of cells that settle in other tissues. Are they incorporated into those tissues, as in the formation of allophenic mice, or do they remain viable but quiescent as cell rests, or do they die?

In studies of progression (see p. 20), especially after selection of tumor cells best able to grow in the ascites, the tumor cells had a great propensity for metastasis. Metastasis was considered as a unit character acquired during progression. Fidler (1973a) has also selected tumors for their propensity to metastasize. Pulmonary metastases of murine B16 melanomas were grown in tissue culture and retransplanted *in vivo*. The pulmonary metastases that resulted were again selected for growth *in vitro* and the cycles repeated. Eventually, tumor lines were obtained with significantly increased incidences of metastases. These cells had surface enzymes (Bosmann *et al.*, 1973) different from their parent stains. Nicolson and Winkelhake (1975) also selected B16 melanoma lines for their ability to metastasize. Although it was claimed that cells were selected for their specific affinities for lung, examination of the data indicates that there was also increased incidence of metastasis in liver, spleen, and kidney. This might suggest that instead of selecting for the ability to metastasize to one organ, the authors were really selecting aggressive cells capable of metastasis. The data do not support the claim of specificity for site of metastasis made by the authors. Greene and Harvey (1964) believe that such specificity exists, but whether or not it does is less important than understanding the characteristics of cells that allow for metastasis. Changes in the electrophoretic mobility of metastasizing cells (Weiss and Hauschka, 1970) and in the enzyme content of their membranes (Bosmann *et al.*, 1973) are of great interest and importance, and again point to cell membranes as being of particular importance in metastasis. Possibly shedding of glycoprotein membrane antigens by tumors may facilitate metastasis (Currie and Alexander, 1974).

Hematogenous metastasis

It is obvious that in order for hematogenous metastasis to occur, tumor cells must be present in the blood. What is not widely realized is that the presence of tumor cells in the blood is not synonomous with metastasis and that metastasis may actually represent a relatively rare consequence of tumor embolization.

Tyzzer demonstrated, early in this century, that palpation or massage of tumors of mice that metastasized in low incidence resulted in an increased incidence of metastasis. This was of concern to surgeons who had to palpate and otherwise manipulate tumors during removal and before the blood supply could be abrogated. A series of clinical studies was done in which the blood of patients with malignant tumors was examined for the presence of circulating tumor cells prior to, during, and after surgery. Tumor cells were found in the peripheral blood in a large percentage of cases, but it wasn't until the study of Engell (1959) that the significance of these observations became clear. Engell examined the venous blood of patients with colonic cancer for circulating tumor cells at the time of surgery, and then he followed the patients to determine whether or not there was a correlation between the presence of circulating tumor cells and survival. He found that approximately one-half of the patients with circulating tumor cells were alive and well from 5 to 9 years after surgery. In other words, the presence of circulating neoplastic cells did not necessarily correlate with a bad prognosis. Crile, Isbister, and Deodhar (1971) support this contention. Other reports have tended to discount the observation and believe that the presence of circulating tumor cells carries a bad prognosis (Roberts *et al.*, 1962). Fidler (1970) found that only 0.1% of circulating cells survived circulation in the blood to form metastasis. It turns out, as might be expected in view of the work of Leighton (1967) on aggregate replication, and the idea that it takes a critical number of cells to form a tissue (Grobstein and Zwilling, 1953), that a small clump of tumor cells might be able to develop into metastasis better than the same number of single cells (Fidler, 1973b).

The usual model used to study hematogenous metastasis has been the intravenous injection of dissociated tumor cells, with subsequent analysis of peripheral blood and organs. Target organs have either been examined histologically for the presence of tumors or portions have been minced up and transplanted into appropriate hosts to determine whether or not metastatic cells were present that were capable of forming a tumor. The disadvantage of the model is that the tumor cells injected may not necessarily be the ones that would normally metastasize. As indicated in Chapter Three, many of the cells in a malignant tumor have differentiated and are benign. There is no way of knowing whether or not the numbers of cells injected correspond with the numbers normally metastasizing. In any event, Dobrossy (1969) has shown that from 15 to 30 minutes after their intravenous administration, most of the malignant cells may be recovered from the circulating blood. This is followed by a latent period of from 3 to 10 days during which few tumor cells are found in the blood and then by a recurrence of tumor cells in the blood. It has been shown by autoradiography that most of the original cells are arrested in the lung and destroyed. The few tumor cells that survive proliferate and are responsible for the secondary or late outgrowth with reseeding of the vasculature.

The main lesson learned from these studies is that, like the peritoneal cavity, the bloodstream is a hostile environment for all but the normal constituents of the blood. The exact reasons for this are not known, but they probably relate to the hemodynamics of blood and the concentration of serum and oxygen. Although most cells in tissue culture require serum supplements of up to 20% for optimal growth, larger concentrations seem to be detrimental. It is conceivable that 100% serum is toxic to many cells. In addition, the blood at a particular time might be carrying toxic metabolites. It is well known that tissue culturists, of necessity, test all serum prior to its use in culture media in order to exclude toxic batches. At any rate, it appears axiomatic that successful metastasis requires cells that are hardy and capable of getting out of the circulation quickly—but what of clumps of cells? Few experiments have been performed to determine if a critical number of cells is necessary to create a microenvironment that would invariably result in metastasis. Successful metastasis probably requires either a shower of single cells or the liberation of aggregates of cells large enough to protect the inner cells from the hazards of embolization and impaction.

Evidence that specific host defense mechanisms inhibit metastasis has not been established with certainty. There are reports that injections of pre-immunized lymphocytes suppress the growth of grafted tumors and there are reports that they do not. Part of this variation of response might be due to the number of sensitized lymphocytes used. In this regard, Fidler (1973b, 1974) obtained a reduction in metastasis when 5,000 sensitized lymphocytes per tumor cell were incubated *in vitro* and then injected into animals. When the ratio of sensitized lymphocytes to tumor cells was 1,000:1, there was an enhancement instead of a reduction of growth. This confirms the observations of Prehn, who believes that normal immune responses may be both stimulatory and inhibitory (Prehn, 1971a and b).

In the discussion of dormancy (p. 19) it was noted that Fisher and Fisher (1967a and b) injected 50 dissociated tumor cells into the portal circulation and that the cells remained dormant. It would have been interesting to learn whether or not metastases would have appeared if the cells had been allowed to aggregate prior to injection. Our guess would be that metastasis would have occurred because the aggregated cells would have formed a critical mass.

Wood (1958) and Wood, Holyoke, and Yardley (1961) used time-lapse cinematography through a window in the ear of a rabbit to determine the sequence of events that followed the arrest of tumor cells in small vessels. Wood's data indicated that tumor cells did not survive long in the vasculature and that the establishment of a metastasis depended upon the rapidity with which the tumor embolus became extravascular. When a cell adhered to endothelium, fibrin and endothelial cells quickly covered it, rendering the tumor cell extravascular. Administration of anticoagulants effectively reduced the incidence of metastasis, emphasizing the importance of a fibrinous coagulum in protecting cells from the adverse effects of the circulation. This micro-clot also appeared to serve as a scaffold for the migration of endothelial cells over the impacted embolus of tumor cells, rendering it extravascular. Wood observed that a leukocyte usually attached to the tumor cell and seemed to precede it through the wall of the blood vessel. The significance of this observation remains to be determined. Gasic *et al.* (1973) have demonstrated a role for platelets interacting with circulating tumor cells in metastasis. Possibly, they cause the microthrombi described by Wood. In the presence of thrombocytopenia there are fewer metastases.

Fisher and Fisher (1967b) have presented evidence that neither fibrinolysis nor anticoagulants affect the incidence of metastasis in their system. Thus, there is some controversy on this matter, and different systems may require different conditions.

Lymphatic metastases

Surgeons, long aware that carcinoma metastasizes first by lymphatics, have done *radical operations* in an effort to extirpate not only primary but also any secondary malignant cells presumably temporarily arrested in lymph nodes. Clinically, the prognosis is best in carcinomas in which there are no metastases to the regional lymph nodes, poorer when one or two nodes are involved, and even worse when more are involved. Patients with multiple node involvement usually have metastases in visceral organs. Zeidman (1957) has shown that cancer cells injected into the lymphatics of experimental animals are filtered out in the peripheral sinuses of lymph nodes, lending logic to the clinical attack. These arrested cells grow, however, and eventually metastasize again (Sugarbaker, Cohen, and Ketcham, 1971). Since there is some evidence that tumor cells may traverse lymph nodes, the nodes may be less of a barrier to the spread of tumor than has been believed (Fisher and Fisher, 1966).

Lymph nodes are known to be a barrier to the spread of infectious processes, but there is little information on whether or not lymph nodes influence the growth of tumors. Zeidman (1965) injected equal numbers of V2 carcinoma cells of the rabbit into the veins or lymphatics and found metastases in 50% and 18% of the animals, respectively. This suggests that the lymphoid system exerted an inhibitory effect upon the growth of neoplastic cells. This may be mechanical or immunologic in nature, but hyperplastic nodes draining tumors seem to favor a better prognosis.

Ascites

Malignant tumors, particularly those of the gastrointestinal tract, invade and often extend through the wall of the organ and seed the abdominal cavity. There they elicit an inflammatory response, circulate, and proliferate as free-growing tumor cells in the resultant exudate. Newly implanted cells elicit additional exudate and slough more cells into the fluid (Goldie, 1956). In the terminal stages of this complication the belly is distended with fluid teeming with rapidly growing, free-floating, extremely undifferentiated malignant cells, a situation in many ways analogous to a suspension tissue culture. Free growth of tumor cells may occur in any body cavity (pleura, etc.).

The method most commonly used to study ascites tumors has been the injection of a mince of tumor into the peritoneal cavity of suitable hosts. Early in these investigations it was apparent, especially to Klein and Klein (1956), that only highly malignant tumors formed an ascites, and that large inocula were required for successful conversion from the solid to the free-floating state. In other words, the intraperitoneal cavity was an inhospitable place for tumor cells. Klein and Klein discovered that some tumors converted immediately, others gradually, and the majority not at all.

In an elegant series of experiments the Kleins demonstrated that conversion to the ascites could be accelerated by alternating intraperitoneal and subcutaneous inoculations. Their observations support the notion that gradual conversion to the ascites was due to selection of tumor cells best able to survive under the conditions of the environment, not adaptation by the overall population of cells injected. In further studies the Kleins injected mixtures of ascites cells and cells from the original solid tumor. If the inoculum contained less than 5% of ascites cells, immediate conversion did not result, and selection through either repeated intraperitoneal inoculation or alternative intraperitoneal and subcutaneous inoculation was necessary to achieve conversion. When a critical number of converted cells was present, they seemed to be able to create an ideal environment for rapid growth. Klein and Klein believe that there is a mutational or metabolic event that contributes to ascites conversion. In this regard, they have studied the composition of ascites fluid and found it to be lacking in essential metabolites and oxygen. Thus, cells capable of growing in this environment must have peculiar metabolic requirements.

Goldie (1956) noted that cells injected intraperitoneally in a mouse quickly collected (via the drainage channels) near the gastric and splenic ligaments. As the cells implanted, these areas became edematous, and between 48 and 96 hours after injection he selected surviving cells by transplanting the edematous tissue to the subcutaneum of other mice. Thus, the resultant tumors had been selected for their capacity to survive for a short period in the intraperitoneum. They were reinjected intraperitoneally, then subcutaneously, and the process repeated until the tumor had been converted to the ascites. By this method, tumors impossible to adapt to the ascites by other methods were converted to ascites tumors.

Stroma

All tumors are supported by a connective tissue stroma. Some have a particularly abundant one (adenocarcinomas of the breast), but others have little or none. Choriocarcinoma is a good example of the latter situation. These growths are characterized by nodules of tumor with necrotic centers and a thin rim of viable tumor cells in contact with the host stroma. Thus, even tumors that are incapable of eliciting a stromal reaction require intimate contact with stroma in order to grow successfully.

Wide experience with heterotransplanted tumors indicates that the heterologous host supplies the tumor with its stroma. Little is known about the controls of this phenomenon, although it has been shown that certain tumor cells resistant to heterotransplantation to the anterior chamber of the guinea pig's eye grow in a much higher percentage of instances if a mixture of embryonic cells is transplanted with the tumor cells. Possibly the embryonic cells help to form a stroma for the tumor.

The role of vascular supply in tumor growth and metastasis has been studied widely by Folkman (1974). Continued growth of a tumor implant requires ingrowth of blood vessels from adjacent non-neoplastic tissue, and in the absence of this vascularization, implants reach a maximum diameter of from 2 to 3 millimeters (Folkman, Cole, and Zimmerman, 1966). Spheroids of tumor cells *in vitro* reach similar diameters and the outer cells continue to live and divide while the central cells degenerate. In either situation, diffusion permits the survival of only a small tumor, but it is insufficient for continued growth.

Wood (1958) observed the rapid onset of endothelial proliferation and capillary bed formation in the immediate vicinity of tumor emboli, and DNA synthesis has been noted in endothelial cells as early as 6 hours after tumor embolization. New capillaries grow into tumors at a rate of up to 1 millimeter per day. In 1968 Greenblatt and Shubik separated tumor implants from the host stroma by a millipore filter and observed stimulation of capillary growth in the host tissue. This suggested the synthesis of a diffusible substance that could cross the millipore filter and stimulate the endothelium of the host.

Folkman *et al.* (1971) have attempted purification of this diffusible factor and have reported a weight of approximately 100,000 daltons. It contains RNA, protein, and lipid. It has been found in various tumors and placenta but, thus far, not in other non-neoplastic tissues. This diffusible factor stimulates DNA synthesis in endothelial cells, but it is not known if the stimulation is specific for these cells.

ELEVEN

Cancer Therapy

Introduction

The ultimate goal of cancer research is to eliminate the disease by prevention of most cases and by treatment of individual cases that break through the public health screen. By analogy with the infectious diseases, prevention would be achieved by identifying causative agents and eradicating them or immunizing against them. We have already outlined the difficulties inherent to these approaches and for practical purposes must focus on developing the means of making early diagnosis, the identification of high risk populations, and therapy of individual cases.

Massive efforts have been made to educate the public to recognize the symptoms of cancer and to teach the methods of examining for skin lesions, lumps in breasts, and so on. Periodic gynecologic examinations and Papanicolaou's smears have led to the earlier diagnosis and improved prognosis of carcinoma of the cervix and, it is hoped, mammography will be similarly beneficial for carcinoma of the breast. When the risk is particularly high, more drastic measures may be indicated. Patients with familial multiple polyposis, for example, have a virtually 100% risk of developing carcinoma of the colon; identification of these individuals and colectomy prior to development of cancer are indicated.

At present, the most frequent approach to cancer in human beings is *post facto*, *i.e.*, treatment of the disease after it has already become established. The objective of the initial therapy is to remove the tumor completely and restore normal function. The excision usually includes a wide rim of normal tissue, sometimes followed by radiation and/or chemotherapy of metastases. Unfortunately, by the time most cancers are discovered, they have already invaded and/or metastasized. When dissemination is evident at the time of presentation, therapy is directed more toward palliation than cure. Thus, despite increased public awareness of the early signs of cancer, despite increasingly early, widespread, and sophisticated diagnostic techniques, and despite half a century of experimental and clinical efforts at therapy, more than half of all patients with cancer of internal organs die of their disease. Many clinicians feel that even if the currently available techniques of diagnosis and therapy were optimally used, this figure would not be significantly altered (Cutler, Myers, and Axtell, 1975).

Carcinoma of the colon is one of the most common cancers of human beings. It often presents as intestinal obstruction, which if untreated is lethal in about a week. This indicates the necessity of immediate initial treatment; the tumor is either removed and normal physiology restored or the patient dies. If the tumor is completely removed, the patient is cured, but if there has been invasion or metastases beyond the limits of resection, the remaining tumor can grow and destroy the host. There is great success with initial therapy. The problem is the therapy for patients with metastatic tumor because the secondary tumors are numerous and widespread and cannot easily be removed. Destruction of metastasis is attempted by chemotherapy and irradiation.

Both irradiation and chemotherapeutic agents are injurious to normal as well as to neoplastic cells. They often have a predilection for proliferating cells, acting at one or more points in the cell cycle. Since cell renewal in marrow and intestine is as rapid as in most tumors, the so-called *cycle cell agents* in therapeutic doses adequate to kill tumor cells also damage normal cells and can cause death. Unfortunately, agents specifically toxic for tumor cells have not been developed. Therapeutic regimens in current use are considered optimal if they provide maximal antitumor effect with tolerable harmful effects in the patient. It is important to note that the basis for selecting a mode of therapy is *empiric*. Previous experience has indicated that for patients similar to the one under consideration with a similar tumor (considering symptomology, stage of disease, etc.), this therapeutic regime has been shown to be *optimal!*

Radiation therapy

DNA replication and mitosis are impaired by radiation—resulting in chromatid breaks and abnormal chromosomes and mitoses (Fox and Lajtha, 1973). The radiosensitivity of a tissue is most closely related to its mitotic index. Thus, a low dose of radiation to a rapidly growing tumor admixed with or adjacent to slowly growing non-neo-plastic tissue would be expected to destroy the tumor while sparing the normal tissue. This differential susceptibility is the theoretical basis for cancer radiotherapy (Friedman, 1974).

Unfortunately, many normal tissues (*e.g.*, bone marrow and gut mucosa) reproduce at a rate at least as fast as commonly occurring neo-plasms, and irradiation does not distinguish between normal and neo-plastic cells. Clinically derived schedules have been devised to mini-mize acute damage to normal cells, but the chronic complications of irradiation (*e.g.*, inflammation, vascular occlusion, and scarring) still occur (White, 1976). Also, tumors that are particularly radiosensitive are often the ones that later become radioresistant.

The effects of radiation on neoplastic as well as non-neoplastic cells have been studied *in vitro*. Synchronized cells *in vitro* have shown dif-ferences in radiosensitivity related to the particular phase in the cell cycle during which the radiation is administered (Karcher and Jentzsch, 1972). The increased radiosensitivity of cells irradiated in G_1 may be due to the impairment of enzymes synthesizing DNA building blocks such as thymidine kinase. Cells irradiated in G_2 undergo chromosomal aberrations. It is hoped that such differential sensitivity might be ap-plicable to cancer therapy (Tolmach, Weiss, and Hopwood, 1971). At-tempts are also being made to find chemicals and other agents that might maximize the differential sensitivities of neoplastic and normal cells to low doses of radiation. These attempts are hampered by incom-plete knowledge of the basic mechanisms by which ionizing radiation damages cells (Ebert, 1973).

Chemotherapy

The use of chemicals to treat cancer began in the late 1940's with Farber's use of methotrexate to treat children with leukemia. As an increasing number of chemicals became available, it became ap-parent that their evaluation required a national systematic effort. Since the 1950's the National Cancer Institute Drug Development Program has tested more than 300,000 compounds; from these have emerged ap-proximately 40 that have shown some degree of effectiveness in the treatment of cancer. In recent years, although the search for new agents has continued, emphasis has also been placed on the more effective application of existing agents. Such efforts have included combining several chemicals, or chemicals with other therapeutic modes, the es-tablishment of sequential schedules, and so on. These empirically derived regimens have resulted in numerous experimental protocols, some of which have markedly improved the prognosis of a few types of cancer. The current status of cancer chemotherapy has been recently reviewed (Burchenal, 1975; Skipper, 1973; Zubrod, 1976).

Chemotherapeutic agents fall into several categories: alkylating agents, antimetabolites, antibiotics, and antimitotics (Chabner *et al.*, 1975a and b). While studying the chemical warfare possibilities of nitrogen mustards during World War II, investigators found that these alkylating agents suppressed lymphosarcoma in mice and in man. Since then they have been used to treat a variety of human cancers. Alkylating agents act by replacing hydrogen in the guanine of DNA. The altered DNA is susceptible to breakage during replication. Cyclophosphamide is an example of an alkylating agent.

The antimetabolites act by interfering with nucleic acid synthesis. Methotrexate binds with dihydrofolate reductase. This enzyme, which normally catalyzes the reduction of dihydrofolate to tetrahydrofolate, is thereby inactivated, and the cell is deprived of the N^5,N^{10}-methylene-tetrahydrofolate needed to synthesize thymidylic acid. Nucleoside analogues such as 5-fluorodeoxyuridine and cytosine arabinoside interfere with the synthesis of the normal nucleotides and thereby cause cell death.

Various antibiotics, many of them synthesized by Streptomyces, bind with DNA and interfere with its replication and/or with transcription. Actinomycin D and adriamycin are among the antibiotics that are currently used.

The antimitotics (vincristine) include vinca alkaloids that interfere with spindle formation. The mechanism of action of the nitrosoureas is unknown, while hydroxyurea appears to act as an inhibitor of ribonucleotide reductase. Glucocorticoids (notably Prednisone) increase the effectiveness of various chemotherapeutic regimes through unknown means.

These agents, plus a few others, and L-asparaginase and hormones (discussed below) form the basic armamentarium of cancer chemotherapists. Their major therapeutic effects are seen in a few (perhaps ten) types of cancer, in which alone, or in conjunction with surgery and/or radiation, they have markedly improved prognosis. In the treatment of uterine choriocarcinoma, a highly malignant, rapidly growing form of cancer, the use of methotrexate or actinomycin D has resulted in a high cure rate, even in the presence of metastases. Similar improvements have been obtained in the treatment of Wilms' tumor, Burkitt's lymphoma, childhood leukemia, advanced Hodgkin's disease, and a few others. Treatment in a few other forms of cancer results in the remission of advanced disease and prolongation of life, but infrequent cures. Unfortunately, in the most common forms of human cancer (lung, kidney, prostate, gastrointestinal tract, breast, uterus) chemotherapy has had minimal effect.

The major drawback to chemotherapeutic agents is not their inability to kill cancer cells, but rather their relative inability to select between neoplastic and non-neoplastic cells. Hence, the toxic and virtually ubiquitous side effects associated with therapeutic doses of these agents, notably, the suppression of tissues with a high rate of cell renewal. Bone marrow suppression results in anemia, thrombocytopenia (lack of platelets), with resultant bleeding, and possibly death. Lack of neutrophils, as well as suppression of lymphoid tissues and immunodeficiency, results in infections, often due to organisms that are not ordinarily pathogenic. In the bowel, failure to renew the mucosa results in ulcerations and other lesions. A variety of other lesions have been described as complications of chemotherapy in many other sites, *e.g.*, in the heart, liver, lung, kidney, and brain.

There is one example of a chemotherapeutic agent that discriminates between neoplastic and non-neoplastic cells. In 1953 Kidd noted an inhibitory effect of guinea pig serum on mouse lymphosarcoma. This effect was due to L-asparaginase, an enzyme that is not present in most other species, including man. Normal cells contain asparagine synthetase and are therefore able to produce their own asparagine; however, some tumor cells lack this enzyme and depend upon exogenous asparagine. The administration of L-asparaginase removes this exogenous source and the asparagine-deficient tumor cells die. L-asparaginase (now obtained from *E. coli*) has been used, with moderate success, in the treatment of human leukemia. Unfortunately, few tumors lack asparagine synthetase.

Hormones play a more or less specific role in the treatment of certain tumors, especially of tumors derived from tissues that are hormone-dependent for their growth and function (see Chapter Two). Withdrawal of hormonal support is widely used in the treatment of breast and prostatic cancer, often resulting in marked improvement of symptomatology and temporary remission, but rarely in cure. Individual tumors respond in varying degrees to hormonal therapy; because of the increasing understanding of the basic mechanisms of target tissue response to hormones, such variability may be explained. For example, human breast cancers have shown variation in the number of estradiol receptor sites, and mouse breast cancers have shown variation in the number of prolactin receptor sites. Hormone-specific receptor sites (or their absence) in such tumors may prove to be therapeutically exploitable (Jensen and DeSombre, 1973). Unfortunately, most tumors of endocrine organs appear to have lost any dependency, presumably as a result of progression.

Chemotherapy is rarely the only form of therapy used in treating cancer, and it is even more unusual for a single chemotherapeutic agent to be used. Exceptions are the use of methotrexate or actinomycin D in uterine choriocarcinoma, cyclophosphamide in Burkitt's lymphoma, and, most recently, streptozotocin, an antibiotic that selectively destroys beta cells, in malignant islet cell tumors of the pancreas. Excellent results, including many cures in spite of the presence of metastases, have been achieved in each of these, but it must be recalled that these are rare tumors in the United States.

More often, chemotherapeutic agents are administered in combination, in the hope that they will act additively or even synergistically. If the agents act in different ways at multiple steps in the cell cycle or on other metabolic pathways, they might be more generally lethal and achieve the ultimate desired effect—that is, elimination of all of the cancer cell population. Finally, giving several different agents might permit a reduced dose of each agent, thereby abating toxic side effects and decreasing the development of resistance to any one of them. An example of this kind of combination chemotherapy is the MOPP (mustard, vincristine, procarbazine, prednisone) regime used in the treatment of advanced Hodgkin's disease—with a notable improvement and apparent cures in many of these patients. Combination chemotherapy has also been curative in many cases of childhood acute lymphoblastic leukemia, formerly rapidly and uniformly fatal (Pinkel, 1971). Unfortunately, we don't know why acute lymphocytic leukemias may respond while other leukemias are refractory to treatment.

The effects of combination chemotherapy, like those of the individual agents themselves, are usually determined empirically, and are sometimes unexpected. Methotrexate, for example, is more effective when given together with vincristine, apparently because in some way vincristine facilitates the entry of methotrexate into cells.

The relationship of mitosis and cell cycle to the effectiveness of *mitotically linked* chemotherapy is well known (Valeriote and Van Putten, 1975). Many of the most effective chemotherapeutic agents act only against cells that are actively dividing, *i.e.*, the growth fraction of a tissue. Ideally, one might recruit all of the tumor stem cells into this vulnerable population and then subject them to these agents. In this way, the hoped for objective, the elimination of the reservoir of stem cells and the eradication of the tumor, might be achieved. The main challenge, using existing therapeutic modes, has been the relatively slow-growing tumors with small growth fractions; these tumors, which are among the most common in man, have been resistant to chemotherapy. There is nothing intrinsic to these tumors, other than their rate of growth, that prevents them from being susceptible to antimitotic agents. The possibility of synchronizing dividing tumor cells and administering an agent that is effective during a particular phase of the cell cycle, thereby concentrating its effect on the tumor, has also been considered (Van Putten, 1974).

One problem in finding agents effective against the common human cancers is that testing in laboratory animals has involved tumors with large growth fractions, *e.g.*, mouse leukemia L1210. A *second-generation* screen, using animal tumors with small growth fractions similar to those of the common human cancers, may be necessary. Two such tumors have been studied, the Lewis lung tumor and B16 melanoma, both are generally unresponsive to most chemotherapeutic agents now in use. *In vitro* screening of potential chemotherapeutic agents has not been very productive.

The problem with chemotherapeutic agents is that they are highly toxic to non-neoplastic as well as neoplastic cells, and the complications and symptomatology caused by these agents may be as severe as those due to the cancer that they are being used to treat (Hall, 1976). The range between the therapeutic and toxic effects of these agents is so narrow that therapists should be congratulated for their successes. Schabel (1975) and others have suggested that therapists should not wait for a cancer to become widespread before using chemotherapy; instead, it is suggested that these agents be used at the time of removal of the tumor. In this way, it is hoped, disseminated disease might be avoided, and the use of lower doses would not result in toxic effects.

The immune response is a highly specific mechanism whereby the host responds to antigens recognized as different from its own. Many experimental and some human cancers bear antigens distinct from those of their non-neoplastic counterparts. This has been the basis for current efforts, both experimental and clinical, at immunotherapy.

Immunization against cancer might be *prophylactic* (an attempt to prevent the development of cancer) or *therapeutic* (an attempt to treat an already established cancer), active, adoptive or passive, and specific or nonspecific. Many recent symposia and review articles discuss this area in detail, *e.g.*, Morton (1972); Mathé and Weiner (1974); Fahey *et al.* (1976); and Bartlett, Kreider, and Purnell (1976).

Immunologic cancer prophylaxis presumes either the identification of infectious oncogenic agents or the isolation of tumor-specific antigens. In the first instance, immunization of the host would protect against infectious oncogenic agents. Animals, for example, can be protected from developing feline leukemia or Marek's disease by vaccination with the causative virus (Purchase, 1976); the development of similar vaccines applicable to man depends upon the isolation of oncogenic viruses in man. There are numerous theoretical and practical impediments to this approach, *e.g.*, vaccination will not work in man if cancer is the result of activation of intracellular proviruses (Mayyasi, Larson, and Ahmed, 1976). In the second instance, it is conceivable that prior immunization to tumor-specific antigens might prevent the development of the corresponding tumors in man, much as prior exposure to tumor extracts prevents the successful transplantation of tumors in experimental animals. Unfortunately, few tumors have common tumor-specific antigens and chemically induced tumors are notable for their antigenic diversity. Both possibilities of immunologic cancer prophylaxis are as yet highly speculative.

Renewed interest in nonspecific immunologic enhancement was stimulated by demonstration that BCG, an attenuated strain of *Mycobacterium bovis* used in tuberculosis prophylaxis, was effective in preventing the successful transplantation of experimental tumors. However, BCG, which stimulates cell-mediated immunocompetence, is ineffective in treating established transplanted tumors. It is currently being evaluated in clinical situations; direct injection of BCG into cutaneous nodules of melanoma results in dramatic regression of the nodules, but there is no effect upon visceral metastases. BCG has been used in treating leukemia, but the results of therapy have been equivocal. Other nonspecific stimulators of host cell-mediated immunity are being investigated.

Attempts have been made specifically to stimulate the host's immune response to a tumor. One approach is to boost the antigenic dose by administering inactivated autologous or homologous tumor cells, which hopefully share tumor antigens. An increased dose of antigen may result in an enhanced immune response. Alternative possibilities are that increased antigen will "soak up" and simply dilute the immune response, diverting it from the tumor, or that it might result in adverse immune responses, *e.g.*, the production of blocking antibodies. At any rate, the results of such attempts, *e.g.*, in treating widespread melanoma and osteosarcoma, have been inconclusive.

Another approach is to alter the host's tumor cells, *in situ* or *in vitro*, thereby modifying and hopefully enhancing their antigenicity. Tumor cells treated with neuraminidase or concanavalin A have enhanced immunogenicity. Neuraminidase, which removes sialic acid residues from cell membranes, is thought to enhance tumor cell antigenicity, perhaps by exposing neoantigens. Con A may enhance the immune response through its ability to bind to cell surfaces and act as a specific T cell mitogen. Cell surface modifiers such as neuraminidase, con A, and proteolytic agents are being used in clinical trials, but results have been equivocal.

The transfer of immunocompetent cells is referred to as *adoptive immunotherapy* and it is said to be specific when the donor bears the same type of humor (with, presumably, the same tumor antigens) as the recipient. The difficulties of adoptive immunotherapy, *e.g.*, in the treatment of leukemia, are outlined by Mathé *et al.* (1973). It may result in histoincompatibility, with rejection of the donor cells or, if the homograft "takes," subsequent "graft vs. host" reaction.

Attempts have been made to circumvent the use of homologous cells by taking the patient's own lymphocytes, exposing them to the host's own tumor cells (or to tumor cells of identical type from other patients) *in vitro*, and reintroducing these now-sensitized, immunocompetent cells back into the original host. An extension of this procedure is to extract and introduce *informational* molecules that direct specific immunologic reactions to tumor antigens.

The basis for immunotherapy rests upon the existence and characterization of tumor-specific antigens and the initiation or enhancement of active and/or passive cytocidal immune responses against the antigen-bearing cells. One basis for caution in the currently optimistic view of immunotherapists is that some *tumor-specific* antigens may in actuality be *stem-cell-specific* and may destroy normal as well as malignant stem cells (Hall, 1974).

Finally, enhancement of differentiation in cancers may be the "new" therapy of cancer. Teratocarcinomas differentiate into tissues indistinguishable from normal. Rarely, neuroblastomas in humans differentiate, in their entirety, into their benign counterparts; they frequently do so *in vitro*. The controls of normal and neoplastic differentiation are unknown. With such knowledge, one could contemplate treating cancers by inducing them to differentiate into benign tissues. Such a therapeutic mode might be much more feasible than trying to identify and eliminate all of the possible etiologic agents of cancer and far less traumatic than surgery, radiation, chemotherapy, and immunotherapy.

Spontaneous regression

Spontaneous regression, although a documented fact, occurs so rarely that it plays no part in the care of cancer patients. In Boyd's monograph, *The Spontaneous Regression of Cancer* (1966) it is stated, "very rarely growth may stop completely and the neoplasm, both primary and secondary, may resolve and eventually disappear. This process is known as regression, and if it occurs in the absence of adequate therapy, it is spoken of as spontaneous." Everson and Cole (1966) reviewed 600 reported cases published since 1912 and considered fewer than 50 to be authentic. To be accepted, the lesion has to be identified by biopsy as a malignant tumor which, with no or inadequate therapy, disappeared over a period of time. The few bona fide examples of spontaneous regression include cases of neuroblastoma, carcinoma of bladder, carcinoma of breast, melanoma, hepatoma, and sarcoma.

In children, regression appears to be related to age (Bolande, 1971). It is known that children under one year of age with neuroblastoma, Wilms' tumor, or yolk sac carcinomas of the testes have a much better prognosis than ones in which the tumors develop later. Interestingly, spontaneous congenital testicular teratocarcinomas of mice grow progressively until puberty; then the vast majority arrest and grow no further, but their differentiated and benign tissues persist. What the factors are in infancy that ameliorate the malignant growth of some tumors is not known, but they might well be associated with growth factors at a particular time in development. With withdrawal of the factor, the dependent tumor would be incapable of maintaining itself. As discussed in Chapter Eight, many growth factors and growth inhibitory factors have been discovered and might possibly play a role in spontaneous regression. It is also conceivable that some of the tumors that spontaneously regress may be dependent upon environmental factors discussed in Chapter Two. Dependent tumors have the capacity for invasion and metastases and they cause death; yet, under appropriate conditions these tumors and their metastases can regress and even disappear.

By definition, spontaneous regression rightfully refers to those situations in which a tumor disappears leaving no trace. Spontaneous differentiation of all of the stem cells in a malignant tumor could lead to cessation of growth and if the differentiated cells were short lived and had no mechanism for renewal, that tumor would disappear. In the case of differentiation of neuroblastoma into ganglioneuroma, however, the ganglion cells are long lived and persist indefinitely. Visfeldt (1963) reported the case of a 6-month-old baby boy who had a neuroblastoma with metastases. The diagnosis was made by biopsy and the baby was treated palliatively. He developed normally but died at age 21 of a bleeding ulcer. At autopsy, ganglioneuroma was present, but no evidence of malignant tumor was found. Unfortunately, spontaneous differentiation of cancers into totally benign tumors is as rare as classical spontaneous regression.

The tenor of the John's Hopkins conference on spontaneous regression of cancer in 1974 attributed immunologic reactions as the most probable mechanism for spontaneous regression of cancer (Lewison, 1976). It is our belief that immune surveillance destroys small numbers of tumor cells carrying potent antigens on their membranes, but it is difficult to envision a large tumor possessed of such weak antigens that it does not elicit lethal immune responses until late in its development when there is maximal tumor load. It is known, however, that in certain cases nonspecific stimulation of the reticuloendothelial system with BCG, for example, can cause regression of tumors.

α_1-lipoproteins of serum and ascites fluid have been considered to be antitumor in nature. However, it is not likely that these molecules, which will in fact kill some tumor cells, could be the postdefense mechanism responsible for spontaneous regression of a huge tumor load late in the course of neoplastic disease. In a recent review Apffel (1976) has described cytotoxic effects of the peroxide-peroxidase halide systems. Again, the question can be raised as to whether or not these are in fact host defense mechanisms specific for tumors, and it is problematical whether or not they could work against a maximal tumor load.

Obviously, insufficient information is available to resolve the mechanism of spontaneous regression. The most important lesson to be learned from spontaneous regression is that it occurs with such extreme rarity that it holds no hope for the individual with a malignant tumor.

Appendix

Hepatoma

Although the suffix "-oma" generally indicates a benign neoplasm, hepatomas are highly malignant neoplasms derived from the liver parenchyma (Figs. A–1 and A–2). They occur in two morphological forms: undifferentiated and nonfunctional with little resemblance to liver tissue or well differentiated and functional with the capacity to synthesize bile. The second form may resemble normal bile duct epithelium and is often difficult to distinguish from adenocarcinomas metastatic to the liver. These two forms of cancer are related because the liver develops from endodermal cells that differentiated into primitive ducts that in turn form hepatic parenchyma.

In parts of Africa (*e.g.*, among the Bantu) hepatomas are among the most common forms of cancer in males, a situation that has been ascribed to the ingestion of aflatoxin B. This mycotoxin is produced by Aspergillus flavus, a frequent contaminant of improperly stored and mouldy grains and peanuts, major staples in the diets of these people. Aflatoxin B has been experimentally confirmed as the cause of outbreaks of hepatomas among trout and turkeys reared on contaminated feeds.

In the United States, hepatomas are rare, except among persons with cirrhosis. Cirrhosis is the outcome of a variety of situations, eventuating in widespread destruction of hepatocytes and disruption of the liver architecture (*eg.*, hepatitis, alcoholism). This is followed by fibrosis and focal or nodular regeneration of hepatocytes. Recently,

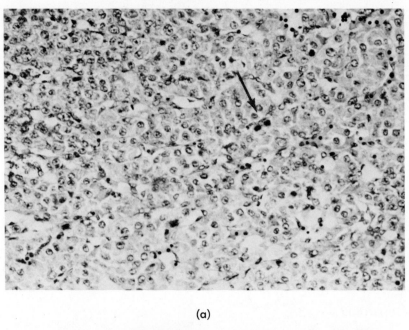

(a)

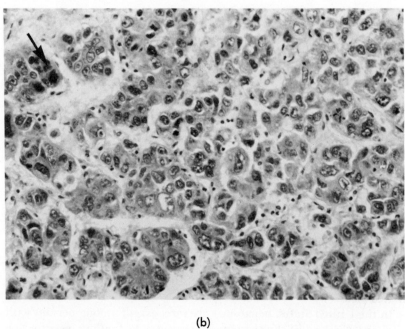

(b)

Fig A–1 (a) Fetal liver (100✕). Even in early development the cells begin to be arranged in cords. A mitotic figure is present (arrow). (b) Hepatocellular carcinoma (100✕). Cells, haphazardly arranged in nests, are pleomorphic, as are their nuclei. A mitotic figure is present (arrow).

there have been increasing, although still rare, reports of hepatocellular neoplasms, variously termed *adenomas* or *low-grade carcinomas,* in women treated with oral contraceptives; these are not associated with cirrhosis.

Hepatomas can be produced by prolonged feeding of various carcinogens to experimental animals. These carcinogens vary extensively in chemical structure, and during the latent period hyperplastic and preneoplastic nodules develop. The cells in these nodules differ little from either normal liver or well differentiated hepatoma cells (Farber, 1973b). A spectrum of differentiation may be found among these tumors, ranging from the highly anaplastic tumors, such as the Novikoff hepatoma (Novikoff, 1957; Pitot, 1960), to well differentiated ones called *minimal deviation hepatomas* (Morris, 1965). Minimal deviation hepatomas appear morphologically and karyotypically similar to adult normal liver and grow slowly, but they have the capacity for metastasis (Morris and Meranze, 1974). They have been extensively studied by Potter (1964), who theorized that since tumors have the capacity for undergoing progression, many of the biochemical alterations associated with neoplasms may not be due to the initial event, but rather to the changes associated with progression. It was his notion that studies of tumors minimally deviated from the normal should reveal the essential biochemical lesions of cancer uncomplicated by the changes of progression.

Many minimal deviation hepatomas have now been studied biochemically, and although many differences are found among the various hepatomas, no consistent pattern that distinguishes them from normal liver and no essential biochemical lesion capable of explaining malignancy have been found.

Hepatomas have been studied cytologically (Ruebner, 1965) and by electron microscopy (Essner and Novikoff, 1962). Daoust and Molnar (1964) have proposed that the cells of hyperplastic nodules are converted to neoplastic cells. The nodules are believed to be regenerative and the result of toxic injury by the carcinogen; it is these regenerative cells that are then initiated by the carcinogen.

Leukemia

Leukemia is a malignant disease of leukopoietic tissues found in many species. The marrow and other reticuloendothelial organs are overgrown by neoplastic cells with replacement of normal blood-forming elements. These malignant leukocytic cells circulate freely in the

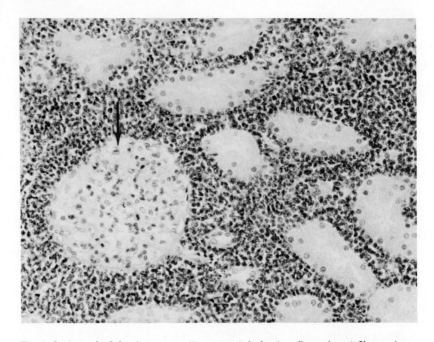

Fig. A–2   Instead of forming masses (*i.e.*, tumors) leukemic cells tend to infiltrate tissues diffusely. In this illustration a dense leukemic infiltrate in the kidney surrounds and separates a glomerulus (arrow) and tubules (40✕). The kidney was large (3 times normal size) and pale (white cell infiltrate).

blood and lymph, and permeate, colonize, and proliferate in the connective tissue stroma of almost all organs (Fig. A–2). This results in enlargement of these organs. Clinical effects of the disease include cachexia and displacement of blood-forming elements in the marrow; the latter results in profound anemia, bleeding tendencies because of *thrombocytenia* (lack of platelets due to replacement of megakaryocytes), and propensity for infection as a result of displacement of nonneoplastic functional leukocytes.

The common types are myeloid (neutrophilic, eosinophilic, basophilic), monocytic, and lymphatic leukemias. Clinically, the disease can occur in acute or chronic forms. The acute form usually present as an acute infection or a bleeding diathesis, if untreated, is fatal in a few weeks or months. The marrow is overgrown by primitive *blast* forms, but the presence of a few partially differentiated cells makes diagnosis of the particular type of leukemia possible. The individuals die of infections, cachexia, or hemorrhage. In the chronic leukemias, blast cells predominate in the marrow, but so many well differentiated leukocytes are released into the circulating blood that the blood may be creamy in color (leukemia means *white blood*). Individuals with chronic leukemia may live for several years, but eventually they die from cachexia or infection.

The overgrowth of the marrow by leukemic cells must not be construed as evidence that the neoplastic cells proliferate more rapidly than the normal cells. Actually, they cycle at a slower rate than normal cells, but too few of their progeny reach a postmitotic state and stop proliferation. The net effect is an enlarged proliferating pool of cells (Astoldi and Mauri, 1953).

The etiology of leukemia has been widely studied. Radiation was implicated when the high incidence of leukemia in dentists and radiologists was correlated with lack of protection from x-rays. Individuals with Down's syndrome have a 10 to 15 times greater propensity for developing leukemia. Nowell and Hungerford (1961) demonstrated an abnormality in the long arm of chromosome 21 in patients with chronic myelocytic leukemia. This chromosome has been named the *Philadelphia chromosome* and results from a translocation of the long arm of chromosome 9. It is the only chromosomal abnormality specific for a particular type of cancer. Leukemia has been extensively studied in animals, particularly with regard to their viral etiology and genetic influences. The viral etiology of leukemia is discussed in Chapter Six.

Neuroblastoma

Neuroblastoma is a malignant tumor that occurs most commonly in the adrenal glands of infants and young children. The tumor is made up of small round cells with scant cytoplasm sometimes arranged in rosette-like patterns; these cells and their configuration resemble primitive neuroblasts [Fig. A–3(a) and (b)]. The tumors are rapidly growing, metastasize widely, particularly to the liver and flat bones of the skull, and, although initially radiosensitive, become radioresistant and cause death from metastasis or cachexia. It is of interest that the prognosis for very young babies with neuroblastoma is better than that for older children, a phenomenon that parallels experience with testicular and Wilms' tumors. The reasons for this are not known.

The undifferentiated tumors just described appear to be one end of a spectrum. At the other end of the spectrum is ganglioneuroma [Fig. A–3(d)], comprised exclusively of benign ganglion cells and Schwann cells. Between these extremes are ganglioneuroblastomas [Fig. A–3(c)].

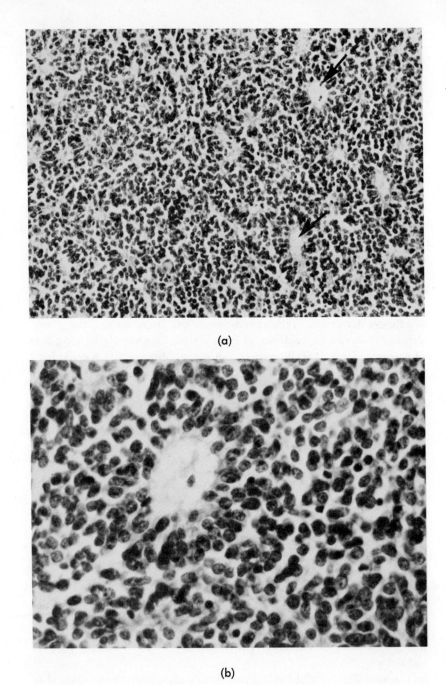

(a)

(b)

Fig. A–3 (a, b) Neuroblastoma (60✕, 300✕). The tumor consists of sheets of small cells containing little cytoplasm and relatively large nuclei. In some areas the neuroblasts are aggregated around a central space to form a rosette (arrows). In actuality, these spaces contain protoplasmic extensions of the neuroblasts.

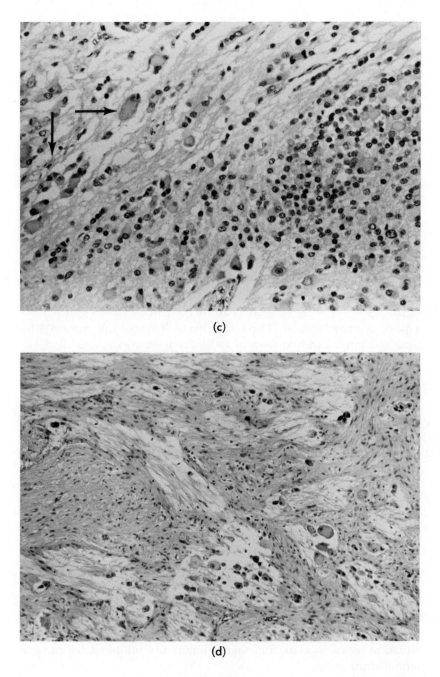

(c)

(d)

Fig. A–3 (cont.) (c) Ganglioneuroblastoma (100×). In this tumor both neuroblasts (lower right) and their more differentiated derivatives, ganglion cells (arrows), are present. The overall appearance begins to resemble that of fetal brain. (d) Ganglioneuroma (100×). Ganglion cells are mixed with nerve; although architecturally haphazard, these components are identical to those of the mature nervous system and neuroblasts are absent.

189

These tumors are comprised of neuroblastoma cells in rosette-like patterns mixed with a few mature ganglion cells [Fig. A—3(c)]. Since neuroblasts normally differentiate into ganglion cells, it would not be surprising if neuroblastomas might occasionally differentiate into ganglioneuromas. There are well documented, although extremely rare, instances in which this has occurred (Cushing and Wolbach, 1927) (see Spontaneous regression, p. 180). This would appear then to be a dramatic clinical situation analogous to that of teratomas in which highly malignant cells spontaneously differentiate into benign cells.

Adenocarcinoma of the breast

Adenocarcinoma of the breast is the most frequent malignant neoplasm causing death in women. The tumors are believed to arise in the epithelium lining the ducts of the breast. They are rare prior to the age of 30. After 30 there is an increase in incidence that reaches a peak at about age 50. There is a lower frequency in women who bear and nurse children than in nulliparous females.

The tumors arise most frequently in the upper-outer quadrant, but they may occur anywhere in the breast. Usually, the tumor presents as a single, rock hard, nontender nodule. If invasion into the muscles behind the breast or the skin overlying the breast has occurred, the nodule feels thick and relatively immovable. Initial spread of the lesion is to the regional lymph nodes. Since the lymphatic drainage of the breast is exceedingly complex, involved nodes may be found in the axillae, supraclavicular area, and thoracic cavity. Metastases to lung and pleura, liver, bone, brain, and adrenal gland are characteristic.

The prognosis is dependent upon the size of the tumor at the time of diagnosis and the stage of the lesion. If the lesion is stage I (no involvement of regional lymph nodes), there is an 80% chance of 5-year survival. However, if there is involvement of the regional lymph nodes, the prognosis drops rapidly. In untreated cases only 20% of patients live 5 years.

Adenocarcinomas of the breast in premenopausal women may be estrogen-dependent and their growth may be retarded by estrogen depletion. Interestingly, in many postmenopausal women the converse situation seems to exist, and many tumors are inhibited by estrogen administration.

The role of endocrine factors in the development of breast cancer has been extensively studied in mice. Geneticists were able to inbreed selectively for strains with a high incidence of spontaneous breast cancer. In such a strain, C$_3$H, 95% of the breeding females develop breast cancer at about 40 weeks of age. Similar animals, maintained as virgins, have an incidence of only 5%. Breast cancer does not occur in male mice, but if the males are castrated and injected with estrogens, the males also develop a significant incidence of breast cancer. Interestingly, men with adenocarcinoma of the prostate treated by castration and estrogen therapy often develop breast enlargement due to ductal hyperplasia and adenocarcinoma of the breast has occurred in a few instances.

Although there are genetic susceptibilities for breast cancer in inbred strains of mice, the overriding factor is sequestration of the mammary tumor virus in animals with a high incidence of adenocarcinoma. This is an RNA virus of B-type that can be vertically transmitted via the milk.

Since there are many similarities between mouse and human breast carcinomas, it is not surprising that a search has been made for the presence of inciting viruses in human beings with adenocarcinoma of the breast. Some of the most interesting studies in this regard are ones undertaken in Parsi women in India, members of a highly inbred religious sect. These women have an exceedingly high incidence of adenocarcinoma of the breast and a B-type virus has been isolated from their milk. Unfortunately, as in all other attempts to demonstrate a viral etiology of human tumors, it is impossible to prove Koch's postulate for this tumor.

Wilms' tumor

Wilms' tumor originates in the kidney of fetuses or infants and has a peak incidence at 3 years. It is twice as common in boys as in girls. The tumor is typically comprised of tubular and glomerular structures resembling those of the early fetal kidney blastema (Fig. A–4). Differentiation along nonrenal lines may also be present, *e.g.*, in the form of smooth or skeletal muscle. Typically, Wilms' tumor presents as a solitary mass within a kidney. It grows rapidly into a large mass, invades adjacent structures, and metastasizes to liver and lungs. Until recently the prognosis has been uniformly poor because these tumors are extremely malignant, recur rapidly after surgery, and although initially radiosensitive, they become radioresistant. With the use of combined surgery, radiotherapy, and chemotherapy, however, 5-year survivals are now achieved in more than 80% of cases. Prognosis is best in very young patients.

Foci that are histologically identical to Wilms' tumor are often incidental findings in the kidneys of abortuses, newborns, and infants. It

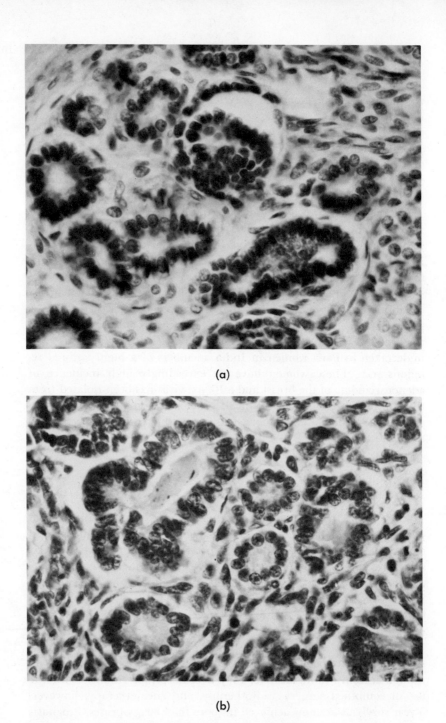

(a)

(b)

Fig. A–4 Normal fetal kidney (400✕). An immature glomerulus is surrounded by immature tubules and stroma. (b) Wilms' tumor (400✕). The tubules and stroma resemble those of the fetal kidney. Structures resembling glomeruli may also be present in Wilms' tumors.

has been suggested that these represent *mini-cancers,* which, for unknown reasons, resolve in the great majority of instances.

A tumor that histologically resembles human Wilms' tumor and that arose spontaneously in the Furth-Columbia rat is being maintained and provides an animal model for the study of this tumor. Similar tumors can be induced in newborn Wistar rats; a single subcutaneous injection of dimethylnitrosamine results in tumors in more than half of the animals months later (Campbell, *et al.*, 1974). Injection of older, though still juvenile, animals requires a larger dose and a longer induction time, results in a lower incidence of tumors, and interestingly, many of the tumors are comprised exclusively of renal epithelial cells. The latter observation suggests that the multipotent stem cell becomes increasingly restricted with aging.

Squamous cell carcinoma

Squamous cell carcinoma may originate in skin, respiratory tract (nose, pharynx, larynx, trachea, bronchi, and lungs), esophagus and anus, vagina and cervix, and other sites. Skin, bronchus, and cervix are the most common locations. Irrespective of site, the tumors are the same morphologically and biologically, and they are made up of stem cells that have a variable capacity for differentiation into squamous epithelium. The squamous cells may be oriented into nodules (called *squamous pearls*) or they may be exceedingly anaplastic with little evidence of squamous differentiation.

The most common site for squamous cell carcinoma is the skin, particularly of the face and hands, that is exposed to environmental carcinogens. Squamous cell carcinomas of the skin are usually less anaplastic than those of the mucous membranes. They usually appear as small, painless nodules that ulcerate and have a crusted granular base surrounded by an elevated border. Although they can invade and metastasize, their location leads to early diagnosis and therapy and they rarely cause death. Those on the lips, oral mucosa, glans penis, vulva, and vagina frequently metastasize to lymph nodes and have a poorer prognosis. There are several agents that play a role in the development of squamous cell carcinoma of the skin, including solar damage (ultraviolet light), chronic irritation, exposure to x-radiation, and exposure to oils, arsenic, and other chemicals.

Squamous cell carcinoma represents approximately 70% of lung cancers. These tumors most frequently arise at bifurcations of large bronchi in men between the ages of 40 and 70 years. It is the most frequent cause of death due to cancer in males. The presence of the tumor interrupts the cilial "house cleaning" mechanism of the lung and predisposes the affected part to inflammation. As the tumor grows, the bronchial lumen becomes narrowed and the resultant obstruction may lead to abscesses in the distal segment of the lung. Thus, lung cancer may present as an intractable pneumonia and the underlying cancer is often not diagnosed until it is well advanced. The tumor invades the bronchi and lung parenchyma and involves the pleura where it may form an exudate with tumor cells growing in the pleural fluid. Blood-borne metastases can occur in any organ and are most frequently found in the liver, adrenal glands, bone, and brain. In untreated cases the most common cause of death is lung abscess and pneumonia. When hematogenous metastases have occurred, the survival rate is markedly reduced. Unfortunately, most cases have metastasized by the time of diagnosis.

The etiology of lung cancer is multifactorial. There is a high incidence in uranium mine workers and in persons exposed to cobalt, nickel, and arsenic. There is a correlation between cigarette smoking and the development of squamous cell carcinoma of the lung. Prolonged smoking results in loss of ciliated respiratory epithelial cells with squamous metaplasia. Squamous cell carcinomas are believed to occur in the metaplastic areas.

In the great majority of cases there is a long history of cigarette smoking. Lung cancers are the most frequent example of ectopic hormone production by tumors. An occasional case, for instance, presents not with the signs and symptoms of a pulmonary mass, but, rather, with a clinical picture of excess ACTH or parathormone production.

A third major type of squamous cell carcinoma occurs in the cervix and is one of the most common forms of cancer in women. Epidemiological studies indicate a relationship between squamous cell carcinoma and age at the time of initial intercourse, frequency of intercourse, and number of partners. It has been suggested that cervical cancer is a venereal disease and an association between herpesviruses and cervical cancer has been reported.

The natural history of carcinoma of the cervix has been extensively studied and has given rise to most of the concepts of preneoplastic states. The earliest event appears to be hyperplasia of the basal cells. Subsequently, the cells develop large hyperchromatic nuclei, pleomorphism, and other abnormalities. This stage is described as *dysplasia*. Some believe that dysplasia progresses to carcinoma *in situ*. A lesion is termed a carcinoma *in situ* when anaplasia is present. Carcinoma *in situ* is found in a younger age group (mean of 35 years) than in invasive carcinoma. In long-term follow-up studies it has been found that patients with carcinoma *in situ* can develop invasive cancer of the cervix.

In experimental studies in which carcinogenic hydrocarbons were painted on the cervix of mice it was found that a series of changes similar to dysplasia and carcinoma *in situ* precedes invasive carcinoma.

The tumor is called an *invasive carcinoma* when the neoplastic cells invade through the basement membrane below the epithelial surface. The prognosis for the patient depends upon the amount and depth of the invading cancer cells at the time of diagnosis.

A typical case history of a patient with carcinoma of the cervix is as follows: The patient was a 43-year-old female who was found to have atypical cells in a Papanicolaou smear (of exfoliated cervical cells) taken at age 31 but was lost to follow-up care. At age 42 a Papanicolaou smear showed anaplastic cells. At that time her cervix contained an ulcerative lesion 1.5 cm in diameter. Biopsy of the lesion showed invasion of squamous cell carcinoma into the uterine wall. The patient was treated with an implant of radioactive gold, but she rapidly developed pain upon defecation and renal failure that led to her death.

At autopsy the tumor was found to have invaded the walls of the large intestine, ureters, and bladder. Ureteral obstruction had resulted in infection and destruction of the kidneys. The sequence of events in this case is typical and the 11 years from the first detection of abnormal cells to the time of tumor detection demonstrates the long period required for the development of this cancer.

Myeloma

Myeloma is an uncommon malignancy that caricatures the development of normal plasma cells (Fig. A–5). Plasma cells are the highly differentiated products of those immune responses characterized by the production of circulating antibodies. Many antigens—drugs, pollen, microorganisms—stimulate the differentiation of lymphocytes (ultimately derived from primitive immunoblasts in the bone marrow) into a homogeneous population or clone of plasma cells. This population produces a similarly homogeneous species of protein (antibody) capable of reacting specifically with the inciting antigen.

In multiple myeloma, numerous foci of neoplastic plasma cell precursors are found in the bone marrow, resulting in the destruction of normal marrow components and adjacent bone. This results in the clinical signs of bone pain, fractures, anemia, and thrombocytopenia, and the characteristic "moth-eaten" appearance of the affected bones on x-ray. Histologically, variable numbers of neoplastic stem cells are mixed with mature appearing plasma cells. Functionally, the neoplastic cells also resemble non-neoplastic plasma cells in that they pro-

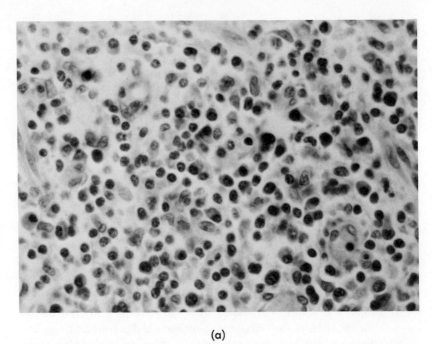

(a)

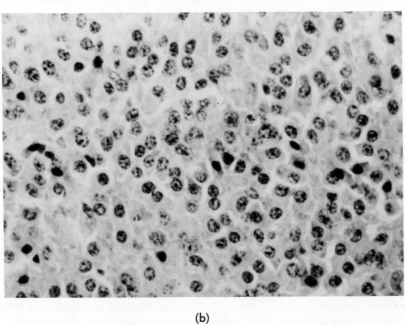

(b)

Fig. A–5 Chronic inflammation. Mature plasma cells predominate in this focus of chronic inflammation (100×). Compare with the appearance of malignant plasma cells in (b). (b) Myeloma (100×). This field is comprised of immature plasma cells. Histologically and ultrastructurally, these resemble non-neoplastic precursors of mature plasma cells.

duce an immunoglobulin that structurally, antigenically, and electrophoretically is very homogeneous. As much as 50 gm of such protein may be produced per day in a patient with myeloma, resulting in a tall, sharp gamma globulin spike in the serum or urine protein electrophoretic pattern. These proteins, termed M or paraproteins collectively, vary from patient to patient. Such variation is consistent with the idea that the tumor arises from a stem cell that is already determined, not only with regard to antibody production, but also as to specificity. Ironically, since these immunoglobulins confer little or no immunocompetence, and since normal means of immunologic defense are overwhelmed by the neoplastic plasma cells, patients with myeloma are susceptible to all sorts of infections.

The urine of patients with myeloma often contains large amounts of protein that demonstrates a unique response to heat. When heated to between 45° and 50°C, the protein precipitates, only to redissolve when the temperature approaches the boiling point. Such Bence Jones proteins have been found to be fragments of entire immunoglobulin light chains, and their small size allows leakage into the urine.

Although myeloma occurs only rarely as a spontaneous tumor in animals, there is an experimentally induced model that has been studied. In 1959 Merwin and Algire reported the experimental induction of myelomas in mice, an unexpected consequence of implanting allogeneic tumor cells in Millipore diffusion chambers in the peritoneal cavities of BALB/c strain animals. Subsequently, it was found that simply implanting bits of plastic or injecting mineral oil or similar irritants resulted in plasma cell tumors in a high percentage of BALB/c or NZB mice. The initial acute inflammatory reaction is followed by granulomata and, in approximately 6 months, by multiple peritoneal tumor nodules that ultimately kill the host. A genetic factor is suggested by the fact that other strains that are similarly treated do not develop tumors. These tumors are transplantable and have been maintained *in vitro*. Through their study, much has been learned about the cell biology and biochemistry of antibody production (Potter, 1968, 1973; Solomon, 1976).

Melanoma

Melanomas (see Fig. A–6) or, more properly, melanocarcinomas, are derived from the same neuroectodermal precursors that give rise to the normal melanocytes of skin and choroid of the eye and to the nevus cells of the benign pigmented mole or nevus. The latter are very common (the average adult having from 15 to 30 nevi) but, although

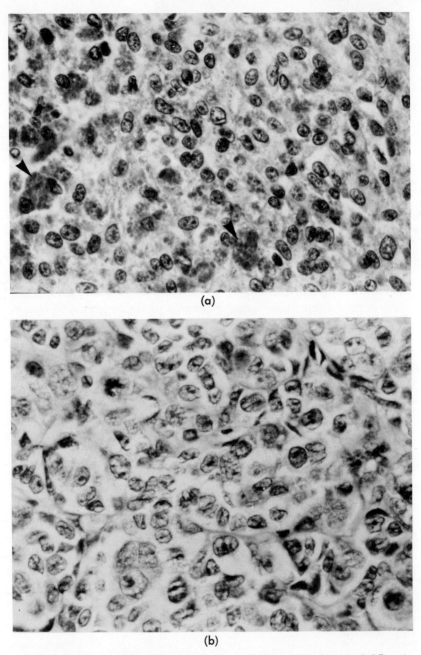

(a)

(b)

Fig. A–6 Melanoma (360×). Pigmentation is the most characteristic feature of differentiation in melanocytes. Its presence or absence is an easily recognized marker. (a) Pigmented (melanotic) melanoma. The cells and their nuclei are relatively uniform and the cytoplasm contains abundant pigment granules (arrows). (b) Nonpigmented (amelanotic) melanoma. The cells and nuclei are larger and pleomorphic. An occasional cell contains a few pigment granules.

the public has been taught to recognize signs of malignant change in nevi, such change is relatively rare, and most melanomas appear to arise *de novo*. Similarly, melanomas have a reputation for being among the most malignant of cancers. In some instances, this reputation is deserved, with a small skin tumor eventuating, within a few months, in widespread metastases to virutally every organ in the body. In other instances, however, the tumor is low grade and the prognosis excellent. This variable behavior is illustrated by the fact that melanomas are among the most frequently documented examples of spontaneous regression (*e.g.*, of metastatic cancer) as well as dormancy (*e.g.*, the recurrence of tumor, in the form of widespread metastases, 20 years after excision of the original lesions). Melanomas are uncommon, tend to occur in young adults as well as the elderly, and are sometimes familial. They are far more common among lightly pigmented than in darkly pigmented persons and races and there is evidence that sunlight plays an etiologic role.

Their production of melanin confers an easily recognizable phenotypic feature [Fig. A–6(a)]. Clinically, nonpigmented melanomas account for approximately 2% of these tumors [Fig. A–6(b)]. Such tumors are comprised exclusively of very immature stem cells. The great majority of melanomas, like other tumors, are comprised of a heterogeneous population of stem cells and their progeny with variable degrees of differentiation—most easily recognized as pigmentation. Not unexpectedly, the amelanotic or least differentiated tumors tend to behave in the most malignant fashion. Pigmentation has been used as a marker of differentiation and has been studied in conjunction with phenotypic features of malignancy in a variety of experimental models, *e.g.*, transplantable tumors, human and animal cells in culture, and cell hybridization systems. Melanomas occur spontaneously in fish, hamsters, and mice. They have been used by Silagi (1967) in studies of differentiation and cancer.

Bibliography

Aaronson, S. A., and G. S. Todaro. 1968. SV40 T antigen induction and transformation in human fibroblast cell strains. Virology 36:254–261.

Abelev, G. I. 1971. Alpha-fetoprotein in ontogenesis and its association with malignant tumors. Adv. Cancer Res. 14:295–358.

Abercrombie, M., and E. J. Ambrose. 1962. The surface properties of cancer cells: A review. Cancer Res. 22:525–548.

Ames, B. N., W. E. Durston, E. Yamasaki, and F. D. Lee. 1973. Carcinogens are mutagens: A simple test system combining liver homogenates for activation and bacteria for detection. Proc. Nat. Acad. Sci., U.S.A. 70:2281–2285.

Ames, B. N., F. D. Lee, and W. E. Durston. 1973. An improved bacterial test system for the detection and classification of mutagens and carcinogens. Proc. Nat. Acad. Sci., U.S.A. 70:782–786.

Anders, F. 1967. Tumour formation in platyfish-swordtail hybrids as a problem of gene regulation. Experientia 23:1–10.

Anderson, D. E. 1970. Genetic varieties of neoplasia. In 23rd M. D. Anderson Conference, pp. 85–105. Baltimore: Williams & Wilkins Co.

Anderson, D. E. 1972. A genetic study of human breast cancer. J. Nat. Cancer Inst. 48:1029–1034.

Anderson, R. E. 1971. Leukemia and related disorders. Hum. Path. 2:505–514.

Apffel, C. A. 1976. Nonimmunological host defenses: A review. Cancer Res. 36:1527–1537.

Arcos, J. C., and M. F. Argus. 1968. Molecular geometry and carcinogenic activity of aromatic compounds: New perspectives. Adv. Cancer Res. 11:305–471.

Atkin, N. B. 1974. Chromosomes in human malignant tumors: A review and assessment. In Chromosomes and Cancer, pp. 375–422. New York: John Wiley & Sons.

Aub, J. C., C. Tieslav, and A. Lankester. 1963. Reaction of normal and tumor cell surfaces to enzymes. I. Wheat-germ lipase and associated mucopolysaccharides. Proc. Nat. Acad. Sci., U.S.A. **50**:613–619.

Aurelian, L., J. D. Strandberg, L. V. Melendez, and L. A. Johnson. 1971. Herpesvirus type 2 isolated from cervical tumor cells grown in tissue culture. Science **174**:704–707.

Axelrad, A. 1969. Genetic and cellular basis of susceptibility or resistance to Friend leukemia virus infection in mice. Canad. Cancer Congr. **8**:313–343.

Bagshawe, K. D. 1973. Recent observations related to the chemotherapy and immunology of gestational choriocarcinoma. Adv. Cancer Res. **18**: 231–263.

Baldwin, R. W. 1973. Immunological aspects of chemical carcinogenesis. Adv. Cancer Res. **18**:1–75.

Baldwin, R. W. 1975. *In vitro* assays of cell-mediated immunity to human solid tumors: Problems of quantitation, specificity, and interpretation. J. Nat. Cancer Inst. **55**:745–748.

Baltimore, D. 1970. Viral RNA-dependent DNA polymerase. Nature **226**: 1209–1211.

Barski, G., and R. Cassingena. 1963. Malignant transformation *in vitro* of cells from C₅₇B1 mouse normal pulmonary tissue. J. Nat. Cancer Inst. **30**:865–883.

Barski, G., and F. Cornefert. 1962. Characteristics of "hybrid"-type cloned cell lines obtained from mixed cultures *in vitro*. J. Nat. Cancer Inst. **28**:801–821.

Barski, G., S. Sorievl, and F. Cornefert. 1961. "Hybrid"-type cells in combined cultures of two different mammalian cell strains. J. Nat. Cancer Inst. **26**:1269–1291.

Bartlett, G. L., J. W. Kreider, and D. M. Purnell. 1976. Immunotherapy of cancer in animals: Models or muddles? J. Nat. Cancer Inst. **56**: 207–210.

Baserga, R. (ed.). 1971. The Cell Cycle and Cancer. New York: Marcel Dekker, Inc.

Baserga, R., and F. Wiebel. 1969. The cell cycle of mammalian cells. Int. Rev. Exp. Path. **7**:1–30.

Bauer, H., R. Kurth, L. Rohrschneider, G. Pauli, R. R. Friis, and H. Gelderblom. 1974. The role of cell surface changes in RNA tumor virus-transformed cells. Cold Spring Harbor Symp. Quant. Biol. **39**: 1181–1185.

Baxt, W., J. W. Yates, H. J. Wallace, J. F. Holland, and S. Spiegelman. 1973. Leukemia-specific sequences in leukocytes of the leukemic member of identical twins. Proc. Nat. Acad. Sci., U.S.A. **70**:2629–2632.

Baxter, J. D., G. G. Rousseau, M. C. Benson, R. L. Garcea, J. Ito, and G. M. Tomkins. 1972. Role of DNA and specific cytoplasmic receptors in glucocorticoid action. Proc. Nat. Acad. Sci., U.S.A. **69**:1892.

Bayreuther, K. 1960. Chromosomes in primary neoplastic growth. Nature **186**:6–9.

Belman, S., and W. Troll. 1974. Phorbol-12-myristate-13-acetate effect on cyclic adenosine 3′,5′-monophosphate levels in mouse skin and inhibition of phorbol-myristate-acetate-promoted tumorigenesis by theophylline. Cancer Res. **34**:3446–3455.

Benjamin, T. L. 1966. Virus specific RNA in cells productively infected or transformed by polyoma virus. J. Mol. Biol. **16**:359–373.

Berenblum, I. 1970a. The study of tumors in animals. *In* General Pathology, 4th ed. (Lord Florey, ed.), pp. 744–780. London: Lloyd-Luke Medical Books, Ltd.

Berenblum, I. 1970b. The epidemiology of cancer. *In* General Pathology, 4th ed. (Lord Florey, ed.), pp. 720–743. London: Lloyd-Luke Medical Books, Ltd.

Berenblum, I. 1970c. The nature of tumour growth. *In* General Pathology, 4th ed. (Lord Florey, ed.), pp. 645–667. London: Lloyd-Luke Medical Books, Ltd.

Berenblum, I., and P. Shubik. 1949. The persistence of latent tumor cells induced in the mouse's skin by a single application of 9:10-dimethyl-1:2-benzathracene. Brit. J. Cancer **3**:384–386.

Berg, J. W. 1976. Nutrition and cancer. Semin. Oncol. **3**:17–23.

Bernhard, W. 1963. Some problems of fine structure in tumor cells. *In* Progress in Experimental Tumor Research, Vol. 3, pp. 1–34. Basel: S. Karger.

Bernhard, W. 1969. Ultrastructure of the cancer cell. *In* Handbook of Molecular Cytology (A. Lima-de-Faria, ed.), pp. 687–715. Amsterdam: North-Holland.

Berwald, Y., and L. Sachs. 1965. *In vitro* transformation of normal cells to tumor cells by carcinogenic hydrocarbons. J. Nat. Cancer Inst. **35**:641–661.

Bishop, J. M., and H. E. Varmus. 1975. The molecular biology of RNA tumor viruses. *In* Cancer (F. Becker, ed.), Vol. 2, pp. 3–48. New York: Plenum Press.

Biskind, M. S., and G. R. Biskind. 1945. Tumor of rat testis produced by heterotransplantation of infantile testis of spleen of adult castrate. Proc. Soc. Exp. Biol. Med. **59**:4–8.

Bittner, J. J. 1939. Relation of nursing to the extra-chromosomal theory of breast cancer in mice. Am. J. Cancer **35**:90–97.

Bittner, J. J. 1947. The causes and control of mammary cancer in mice. Harvey Lect. **42**:221–246.

Blot, W. J., and J. F. Fraumeni, Jr. 1975. Arsenical air pollution and lung cancer. Lancet **2**:142–144.

Bolande, R. P. 1971. Benignity of neonatal tumors and concept of cancer repression in early life. Am. J. Dis. Child. **122**:12–14.

Bonneville, K. 1950. New facts on mesoderm formation and proamnion derivatives in the normal mouse embryo. J. Morph. **86**:495–546.

Borek, E., and S. J. Kerr. 1972. Atypical transfer RNAs and their origin in neoplastic cells. Adv. Cancer Res. **15**:163–190.

Bosmann, H. B., G. F. Bieber, A. E. Brown, K. R. Case, D. M. Gersten,

T. W. Kimmerer, and A. Lione. 1973. Biochemical parameters correlated with tumor cell implantation. Nature 246:487–489.

Bosmann, H. B., A. Hagopian, and E. H. Eylar. 1968. Membrane glycoprotein biosynthesis: Changes in levels of glycosyl transferases in fibroblasts transformed by oncogenic viruses. J. Cell Physiol. 72:81–88.

Bosombrio, M. A. 1970. Search for common antigenicities among twenty-five sarcomas induced by methylcholanthrene. Cancer Res. 30:2458–2462.

Bottomley, R. H., A. L. Trainer, and P. T. Condit. 1971. Chromosome studies in a "cancer family." Cancer 28:519–528.

Boutwell, R. K. 1974. The function and mechanism of promoters of carcinogenesis. CRC Crit. Rev. Toxicol. 2:419–443.

Boveri, T. 1929. The Origin of Malignant Tumors (M. Boveri, ed.). Baltimore: Williams & Wilkins Co.

Boyd, W. 1966. The Spontaneous Regression of Cancer. Springfield, Ill.: Charles C. Thomas.

Boyland, E. 1969. The correlation of experimental carcinogenesis and cancer in man. Progr. Exp. Tumor Res. 11:222–234.

Bradley, T. R., and D. Metcalf. 1966. The growth of mouse bone marrow cells *in vitro*. Aust. J. Biol. Med. Sci. 44:287–300.

Braun, A. C. 1956. The activation of two-growth-substance systems accompanying the conversion of normal to tumor cells in crown gall. Cancer Res. 16:53–56.

Braun, A. C. 1969. The Cancer Problem; A Critical Analysis and Modern Synthesis. New York: Columbia University Press.

Braun, A. C. 1972. The usefulness of plant tumor systems for studying the basic cellular mechanisms that underlie neoplastic growth generally. *In* Cell Differentiation (R. Harris, P. Allin, and D. Viza, eds.), pp. 115–118. Copenhagen: Munksgaard.

Bregula, U., G. Klein, and H. Harris. 1971. The analysis of malignancy by fusion. II. Hybrids between Ehrlich cells and normal diploid cells. J. Cell Sci. 8:673–680.

Bresnick, E. 1974. Nucleotides: Biosynthesis, inhibition of synthesis and development of resistance to inhibitors. *In* The Molecular Biology of Cancer (H. Busch, ed.), pp. 277–307. New York: Academic Press.

Bridges, B. A. 1976. Short-term screening tests for carcinogens. Nature 261: 195–200.

Briggs, R., and T. J. King. 1952. Transplantation of living nuclei from blastula cells into enucleated frogs' eggs. Proc. Nat. Acad. Sci., U.S.A. 38:455–463.

Briggs, R., and T. J. King. 1953. Factors affecting the transplantability of frog embryonic cells. J. Exp. Zool. 122:485–506.

Brinkley, B. R. G. M. Fuller, and D. P. Highfield. 1975. Cytoplasmic microtubules in normal and transformed cells in culture: Analysis by tubulin antibody immunofluorescence. Proc. Nat. Acad. Sci., U.S.A. 72:4981–4985.

Brinster, R. L. 1975. Can teratocarcinoma cells colonize the mouse embryo?

In: Teratomas and Differentiation (Sherman, M. I. and D. Solter, eds.), pp. 51–58. New York: Academic Press.

Brodey, R. S. 1970. Canine and feline neoplasia. Adv. Vet. Sci. 14:309–354.

Broome, J. D. 1961. Evidence that the L-asparginase activity of guinea pig serum is responsible for its antilymphoma effects. Nature 191: 1114–1115.

Buck, C. A., M. C. Glick, and L. Warren. 1970. A comparative study of glycoproteins from the surface of control and Rous sarcoma virus transformed hamster cells. Biochemistry 9:4567–4576.

Bullough, W. D. and E. B. Laurence. 1959–60. The control of epidermal mitotic activity in the mouse. Proc. Roy. Soc. London, Series B 151:517–536.

Bullough, W. S. and E. B. Laurence. 1968. Control of mitosis in rabbit VX2 epidermal tumors by means of the epidermal chalone. Eur. J. Cancer 4:587–594.

Burchenal, J. H. 1975. From wild fowl to stalking horses: Alchemy in chemotherapy. Cancer 35:1121–1135.

Burger, M. M. 1970. Proteolytic enzymes initiating cell division and escape from contact inhibition of growth. Nature 227:170–171.

Burkitt, D. 1958. A sarcoma involving the jaws in African children. Br. J. Surg. 46:218–223.

Burkitt, D. P. 1973. Burkitt's lymphoma. *In*: Cancer in Childhood (J. O. Godden, ed.), pp. 209–215. New York: Plenum Press.

Burnet, F. M. 1971. Immunological surveillance in neoplasia. Transplant. Rev. 7:3–25.

Busch, H. 1974. Messenger RNA and other high molecular weight RNA. *In*: The Molecular Biology of Cancer, pp. 188–239. New York: Academic Press.

Butel, J. S., S. S. Tevethia, and J. L. Melnick. 1972. Oncogenicity and cell transformation by papovavirus SV40: The role of the viral genome. Adv. Cancer Res. 15:1–55.

Cairnie, A. B., P. K. Lala, and D. G. Osmond. 1976. Stem Cells of Renewing Cell Populations. New York: Academic Press.

Campbell, J. S., G. S. Wiberg, H. C. Grice, and P. Lou. 1974. Stromal nephromas and renal cell tumors in suckling and weaned rats. Cancer Res. 34:2399–2404.

Carter, L. J. 1974. Cancer and the environment. I. A creaky system grinds on. Science 186:239–242.

Caspersson, T., G. Lomakka, and L. Zech. 1971. The 24 fluorescence patterns of the human metaphase chromosomes: Distinguishing characteristics and variability. Hereditas 67:89–102.

Casto, B. C., W. J. Piecznski, and J. A. DiPaolo. 1974. Enhancement of adenovirus transformation by treatment of hamster embryo cells with diverse chemical carcinogens. Cancer Res. 34:72-78.

Chabner, B. A., C. E. Myers, C. N. Coleman, and D. G. Johns. 1975a. The clinical pharmacology of antineoplastic agents. Part I. N. Eng. J. Med. 292:1107–1113.

Chabner, B. A., C. E. Myers, C. N. Coleman, and D. G. Johns. 1975b. The clinical pharmacology of antineoplastic agents. Part II. N. Eng. J. Med. **292**:1159–1168.

Chang, J. P., and C. W. Gibley, Jr. 1968. Ultrastructure of tumor cells during mitosis. Cancer Res. **28**:521–525.

Chang, W. W. L., and C. P. Leblond. 1971. Renewal of the epithelium in the descending colon of the mouse. Am. J. Anat. **131**:73–99.

Chen, T. T., and C. Heidelberger. 1969. Quantitative studies on the malignant transformation of mouse prostate cells by carcinogenic hydrocarbons *in vitro*. Int. J. Cancer **4**:166–178.

Chervenick, P. A., and A. F. LoBuglio. 1972. Human blood monocytes: Stimulators of granulocyte and mononuclear colony formation *in vitro*. Science **178**:164.

Christman, J. K., S. Silagi, E. W. Newcomb, S. C. Silverstein, and G. Acs. 1975. Correlated suppression by 5-bromodeoxy-uridine of tumorigenicity and plasminogen activator in mouse melanoma cells. Proc. Nat. Acad. Sci., U.S.A. **72**:47–50.

Church, R. B., S. W. Luther, and B. J. McCarthy. 1969. RNA synthesis in Taper hepatoma and mouse liver. Biochem. Biophys. acta. **190**:30–37.

Clarkson, B., and R. Baserga (eds.). 1974. Control of proliferation in animal cells. Cold Spring Harbor Conference on Cell Proliferation, Vol. 1, pp. 1–1015. Cold Spring Harbor, New York: Cold Spring Harbor Laboratory.

Coggin, J. H., Jr., K. R. Ambrose, B. B. Bellomy, and N. G. Anderson. 1971. Tumor immunity in hamsters immunized with fetal tissues. J. Immunol. **170**:526–533.

Coggin, J. H., and N. G. Anderson. 1972. Phase-specific autoantigens (fetal) in model tumor systems. *In* Embryonic and Fetal Antigens in Cancer. (N. G. Anderson, J. H. Coggin, Jr., E. Cole, and J. W. Holleman, eds.), pp. 91–104. Proceedings of the Second Conference, Oak Ridge National Laboratory, February 14, 1972.

Coggin, J. H., Jr., and N. G. Anderson. 1974. Cancer, differentiation and embryonic antigens: Some central problems. Adv. Cancer Res. **19**:105–165.

Coman, D. R. 1944. Decreased mutual adhesiveness, a property of cells from squamous cell carcinomas. Cancer Res. **4**:625–629.

Coman, D. R. 1953. Mechanisms responsible for the origin and distribution of blood-borne tumor metastases: A review. Cancer Res. **13**:397–404.

Comings, D. E. 1973. A general theory of carcinogenesis. Proc. Nat. Acad. Sci., U.S.A. **70**:3324–3328.

Coon, H. G., and R. D. Cahn. 1966. Differentiation *in vitro*: Effects of Sephadex fractions of chick embryo extract. Science **153**:1116–1119.

Cornell, R. 1969. Spontaneous neoplastic transformation *in vitro*: Uutrastructure of transformed cell strains and tumors produced by injection of cell strains. J. Nat. Cancer Inst. **43**:891–906.

Crile, G., W. Isbister, and S. D. Deodhar. 1971. Lack of correlation between the presence of circulating tumor cells and the development of pulmonary metastases. Cancer 28:655–656.

Croce, C. M., and H. Koprowski. 1973. Enucleation of cells made simple and rescue of SV40 by enucleated cells made even simpler. Virology 51: 227–229.

Cuatico, W., J.-R. Cho, and S. Spiegelman. 1974. Evidence of particle-associated RNA-directed DNA polymerase and high MW RNA in human gastrointestinal and lung malignancies. Proc. Nat. Acad. Sci., U.S.A. 71:3304–3308.

Culp, L. A., W. J. Grimes and P. H. Black. 1971. Contact-inhibited revertant cell lines isolated from Simian virus 40-transformed cells. I. Biologic, virologic, and chemical properties. J. Cell Biol. 50:682–690.

Cunningham, D. D., C. R. Trash, and R. D. Glynn. 1974. Initiation of division of density-inhibited fibroblasts by glucocorticoids. *In* Control of Proliferation in Animal Cells (B. Clarkson and R. Beserga, eds.). Cold Spring Harbor, New York: Cold Spring Harbor Laboratory.

Currie, G. A., and P. Alexander. 1974. Spontaneous shedding of TSTA by viable sarcoma cells: Its possible role in facilitating metastatic spread. Brit. J. Cancer 29:72–75.

Curtis, H. J. 1965. Formal discussion of: Somatic mutations and carcinogenesis. Cancer Res. 25:1305–1310.

Cushing, H., and S. B. Wolbach. 1927. The transformation of a malignant paravertebral sympathicoblastoma into a benign ganglioneuroma. Am. J. Path. 3:203–216.

Cutler, S. J., M. H. Myers, and L. M. Axtell. 1975. Trends in cancer patient survival rates. Progr. Clin. Cancer 6:1–13.

Damjanov, I., D. Solter, M. Belicza, and N. Skreb. 1971. Teratomas obtained through extrauterine growth of seven-day mouse embryos. J. Nat. Cancer Inst. 46:471–475.

Daoust, R., and F. Molnar. 1964. Cellular populations and mitotic activity in rat liver parenchyma during azo dye carcinogenesis. Cancer Res. 24:1898–1909.

Dawe, C. J. 1972. Epithelial-mesenchymal interactions in relation to the genesis of polyoma virus-induced tumours of mouse salivary gland. *In* Tissue Interactions in Carcinogenesis (D. Tarin, ed.). New York: Academic Press.

Defendi, V. 1963. Effect of SV_{40} virus immunization on growth of transplantable SV_{40} and polyoma virus tumors in hamsters. Proc. Soc. Exp. Biol. Med. 113:12–16.

Defendi, V., and J. M. Lehman. 1965. Transformation of hamster embryo cells *in vitro* by polyoma virus: Morphological, karyological, immunological, and transplantation characteristics. J. Cell Comp. Physiol. 66:351–410.

Defendi, V., J. Lehman, and P. Kraemer. 1963. "Morphologically normal" hamster cells with malignant properties. Virology 19:592–598.

DeGrouchy, J., and C. DeNava. 1968. A chromosomal theory of carcino-genesis. Ann. Int. Med. **69**:381–391.

deLustig, E. S., and E. L. Matos. 1972. Regeneration and carcino-genesis: Carcinogenesis in regenerating tadpole tails of *Bufo Arenarum. In* Cell Differentiation (R. Harris, P. Allin, and D. Viza, eds.), pp. 124–130. Copenhagen: Munksgaard.

DePetris, S., M. C. Raff, and L. Mallucci. 1973. Ligand-induced redistribution of concanavalin A receptors on normal, trypsinized and transformed fibroblasts. Nature, New Biol. **244**:275–278.

Dewey, M. M., and L. Barr. 1962. Intercellular connection between smooth muscle cells: the nexus. Science **137**:670–672.

DiBerardino, M. A., and N. Hoffner. 1971. Development and chromosomal constitution of nuclear-transplants derived from male germ cells. J. Exptl. Zoology **176**:61–72.

DiBerardino, M. A., and T. J. King. 1967. Development and cellular differen-tiation of neural nuclear-transplants of known karyotypes. Dev. Biol. **15**:102–128.

Dinman, B. D. 1974. The Nature of Occupational Cancer. Springfield, Ill.: C. C. Thomas.

DiPaolo, J. A. 1974. Quantitative aspects of *in vitro* chemical carcinogenesis. *In* Chemical Carcinogenesis, Part B (Ts'o, P.O.P., and J. A. DiPaolo, eds.), pp. 443–455. New York: Dekker.

DiPaolo, J. A., and P. J. Donovan. 1967. Properties of Syrian hamster cells transformed in the presence of carcinogenic hydrocarbons. Exptl. Cell Res. **48**:361–377.

Dixon, F. J., and R. A. Moore. 1953. Testicular tumors—a clinicopathologic study. Cancer **6**:427–454.

Dmochowski, L. 1954. Progress in mammalian genetics and cancer. J. Nat. Cancer Inst. **15**:785–787.

Dmochowski, L. 1966. Present status of viruses as causative agents in cancer. Southern Med. Bull. **54**:65.

Dmochowski, L., T. Yumoto, C. E. Grey, R. L. Hales, P. L. Langford, H. G. Taylor, E. J. Freireich, C C. Shullenberger, J. A. Shively, and C. D. Howe. 1967. Electron microscopic studies of human leukemia and lymphoma. Cancer **20**:760–777.

Dobrossy, L. 1969. Malignant cells in the peripheral blood and their signifi-cance in metastasis formation., Acta Cytologica **13**:395–398.

Doniach, I., and S. R. Pelc. 1950. Autoradiograph technique. Brit. J. Radiol. **23**:184–192.

Dorsey, J. K., and S. Roth. 1973. Adhesive specificity in normal and trans-formed mouse fibroblasts. Develop. Biol. **33**:249–256.

Dulbecco, R., L. H. Hartwell, and M. Vogt. 1965. Induction of cellular DNA synthesis by polyoma virus. Proc. Nat. Acad. Sci., U.S.A. **53**:403–410.

Duran-Reynals, F. 1941. Age susceptibility of ducks to the virus of the Rous sarcoma and variation of the virus in ducks. Science **93**:501–502.

Duran-Reynals, F. 1952. Studies on combined effects of fowl pox virus and methylcholanthrene in chickens. Ann. New York Acad. Sci. 53: 977–991.

Eagle, H. 1974. Some effects of environmental pH on cellular metabolism and function. *In* Control of Animal Cell Proliferation (B. Clarkson and R. Baserga, eds.), Cold Spring Harbor Conference on Cell Proliferation, pp. 1–11. Cold Spring Harbor, New York: Cold Spring Harbor Laboratory.

Eagle, H., and G. E. Foley. 1956. The cytotoxic action of carcinolytic agents in tissue culture. Am. J. Med. 21:739.

Earle, W. R. 1943. Production of malignancy *in vitro*. IV. The mouse fibroblast cultures and changes seen in the living cells. J. Nat. Cancer Inst. 4:165–212.

Earle, W. R., and A. Nettleship. 1943. Production of malignancy *in vitro*. V. Results of injections of cultures into mice. J. Nat. Cancer Inst. 4:213–227.

Earle, W. R., and C. Voegtlin. 1940. A further study of the mode of action of methylcholanthrene on normal tissue culture. U.S. Public Health Rep. 55:303–322.

Ebert, H. 1973. Molecular radiobiology. Br. Med. Bull. 29:12–15.

Eckhart, W. R., R. Dulbecco, and M. M. Burger. 1971. Temperature-dependent surface changes in cells infected or transformed by a thermosensitive mutant of polyoma virus. Proc. Nat. Acad. Sci., U.S.A. 68:283–286.

Eddy, B. E., G. S. Borman, W. H. Berkeley, and R. D. Young. 1961. Tumors induced in hamsters by injection of Rhesus monkey kidney cell extracts. Proc. Soc. Exp. Biol. Med. 107:191–197.

Eddy, B. E., S. E. Stewart, and W. Berkeley. 1958. Cytopathogenicity in tissue cultures by a tumor virus from mice. Proc. Soc. Exp. Biol. Med. 98:848–851.

Ellerman, V. and O. Bang. 1908. Centr. Bakt. 1. Abt. Orig. 46:595.

Engell, H. C. 1959. Cancer cells in the blood: A 5–9 year followup study. Ann. Surg. 149:457–461.

Ephrussi, B. 1972. Hybridization of Somatic Cells. Princeton, N.J.: Princeton University Press.

Epstein, M. A., B. G. Achong, and Y. M. Barr. 1964. Virus particles in cultured lymphoblasts from Burkitt's lymphoma. Lancet 1:703–705.

Erikson, E., and R. L. Erikson. 1976. The origin of 7S RNA in virions of avian sarcoma viruses. Virology 72:518–522.

Erikson, R. L. 1969. Studies on the RNA from avian myeloblastosis virus. Virology 37:124–131.

Essner, E., and A. B. Novikoff. 1962. Cytological studies on two functional hepatomas. J. Cell Biol. 15:289–312.

Estensen, R. D., J. W. Hadden, E. M. Hadden, F. Touraine, J.-L. Touraine, M. K. Haddox, and N. D. Goldberg. 1974. Phorbol myristate acetate: Effects of a tumor promoter on intracellular cyclic GMP in mouse fibroblasts and as a mitogen on human lymphocytes. *In* Control of

Proliferation in Animal Cells (B. Clarkson and R. Baserga, eds.), Cold Spring Harbor Conference on Cell Proliferation. Cold Spring Harbor, New York: Cold Spring Harbor Laboratory.

Everson, T. Z., and W. H. Cole. 1966. Spontaneous Regression of Cancer. Philadelphia: W. B. Saunders Co.

Ewing, J. 1940. Neoplastic Diseases, 4th ed. Philadelphia: W. B. Saunders Co.

Fahey, J. L., S. Brosman, R. C. Ossorio, C. O'Toole, and J. Zighelboim. 1976. Immunotherapy and human tumor immunology. Ann. Int. Med. 84:454–465.

Farber, E. 1973a. Carcinogenesis-cellular evolution as a unifying thread. Presidential Address. Cancer Res. 33:2537–2550.

Farber, E. 1973b. Hyperplastic liver nodules. *In* Methods in Cancer Research (H. Busch, ed.), Vol. VII, pp. 345–375. New York: Academic Press.

Fell, A. B., and E. Mellanby. 1953. Metaplasia produced in cultures of chick ectoderm of high vitamin A. J. Physiol. 119:470–488.

Fidler, I. J. 1970. Metastasis: Quantitative analysis of distribution and fate of tumor emboli labeled with ^{125}I-5-iodo-2'-deoxyuridine. J. Nat. Cancer Inst. 45:773–782.

Fidler, I. J. 1973a. Selection of successive tumor lines for metastasis. Nature New Biol. 242:148–149.

Fidler, I. J. 1973b. The relationship of embolic homogeneity, number, size and viability to the incidence of experimental metastasis. Europ. J. Cancer 9:223–227.

Fidler, I. J. 1974. Immune stimulation-inhibition of experimental cancer metastasis. Cancer Res. 34:491–498.

Figueroa, W. G., R. Raszkowski, and W. Weiss. 1973. Lung cancer in chloromethyl methyl ether workers. N. Eng. J. Med. 288:1096–1097.

Finch, B. W., and B. Ephrussi. 1967. Retention of multiple developmental potentialities by cells of mouse testicular teratocarcinoma during prolonged cultures *in vitro* and their extinction upon hybridization with cells of permanent lines. Proc. Nat. Acad. Sci., U.S.A. 57: 615–621.

Fisher, B., and E. R. Fisher. 1959. Experimental evidence in support of the dormant tumor cell. Science 130:918–919.

Fisher, B., and E. R. Fisher. 1966. Transmigration of lymph nodes by tumor cells. Science 152:1397–1398.

Fisher, B., and E. R. Fisher. 1967a. Metastases of cancer cells. *In* Methods in Cancer Research (H. Busch, ed.), Vol. 1, pp. 243–286. New York: Academic Press.

Fisher, B., and E. R. Fisher. 1967b. Anticoagulants and tumor cell lodgment. Cancer Res. 27:421–425.

Fisher, E. R., and W. Jeffrey. 1965. Ultrastructure of human normal and neoplastic prostate. Am. J. Clin. Path. 44:119–134.

Fishman, P. H., V. W. McFarland, P. T. Mora, and R. O. Brady. 1972. Ganglioside biosynthesis in mouse cells: Glycosyltransferase activities

in normally and virally transformed lines. Biochem. Biophys. Res. Commun. **48**:48–57.

Fishman, W. H., and R. M. Singer. 1975. Ectopic isoenzymes: Expression of embryonic genes in neoplasia. *In* Cancer (F. Becker, ed.), Vol. 3, pp. 57–80. New York: Plenum Press.

Foley, E. J. 1953. Antigenic properties of methylcholanthrene-induced tumors in mice of the strain of origin. Cancer Res. **13**:835–837.

Folkman, J. 1974. Tumor angiogenesis. Adv. Cancer Res. **19**:331–358.

Folkman, J., P. Cole, and S. Zimmerman. 1966. Tumor behavior in isolated perfused organs: *In vitro* growth and metastases of biopsy material in rabbit thyroid and canine intestinal segment. Ann. Surg. **164**: 491–502.

Folkman, J., E. Merler, C. Abernathy, and G. Williams. 1971. Isolation of a tumor factor responsible for angiogenesis. J. Exp. Med. **133**:275–288.

Fosger, J. M., D. D. Choie, and E. C. Friedberg. 1976. Cancer Res. **36**:258.

Foulds, L. 1956. The histologic analysis of mammary tumors of mice. II. The histology of responsiveness and progression. The origins of tumors. J. Nat. Cancer Inst. **17**:713–753.

Foulds, L. 1969. Neoplastic Development, Vol. 1. New York: Academic Press.

Fox, B. W., and L. G. Lajtha. 1973. Radiation damage and repair phenomena. Br. Med. Bull. **29**:16–22.

Frankfurt, O. S. 1967. Mitotic cycles and cell differentiation in squamous cell carcinomas. Int. J. Cancer **2**:304–310.

Franks, L. M., and P. D. Wilson. 1970. "Spontaneous" neoplastic transformation *in vitro*: The ultrastructure of the tissue culture cell. Eur. J. Cancer **6**:517–523.

Fraumeni, J. F., Jr. (ed.). 1975. Persons at High Risk of Cancer—An Approach to Cancer Etiology and Control, pp. 1–544. New York: Academic Press.

Freeman, A. E., R. V. Gilden, M. L. Vernon, R. G. Wolford, P. E. Hugunin, and R. J. Huebner. 1973. 5-Bromo-2-deoxyuridine potentiation of transformation of rat embryo cells induced *in vitro* by 3-methylcholanthrene induction of rat leukemia virus gs antigen in transformed cells. Proc. Nat. Acad. Sci., U.S.A. **70**:2415–2419.

Freeman, A. E., P. J. Price, H. J. Igel, J. C. Young, J. M. Maryak, and R. J. Huebner. 1970. Morphological transformation of rat embryo cells induced by diethylnitrosamine and murine leukemia viruses. J. Nat. Cancer Inst. **44**:65–78.

Friedman, M. (ed.). 1974. The Biological and Clinical Basis of Radiosensitivity. Springfield, Ill.: Charles C. Thomas.

Friedmann, I., and E. S. Bird. 1969. Electron-microscope investigation of experimental rhabdomyosarcoma. J. Path. **97**:375–382.

Friedwald, W. F., and P. Rous. 1950. The pathogenesis of deferred cancer: A study of the after effects of methylcholanthrene upon rabbit skin. J. Exp. Med. **91**:459–484.

Friend, C., H. D. Preisler, and W. Scher. 1974. Studies on the control of

differentiation of murine virus-induced erythroleukemic cells. *In* Current Topics in Developmental Biology (A. Monroy and A. A. Moscona, eds.), Vol. 8, pp. 81–101. New York: Academic Press.

Furth, J. 1953. Conditioned and autonomous neoplasms. A review. Cancer Res. **13**:477–492.

Furth, J. 1967–1968. Pituitary cybernetics and neoplasia. Harvey Lect., Series 62, pp. 47–71.

Gallagher, R. E., and R. C. Gallo. 1975. Type C RNA tumor virus isolated from cultured human acute myelogenous leukemia cells. Science **187**:350–353.

Gasic, G. J., T. B. Gasic, N. Galanti, T. Johnson, and S. Murphy. 1973. Platelet-tumor-cell interaction in mice. The role of platelets in the spread of malignant disease. Int. J. Cancer **11**:704–718.

Gelboin, H. V., N. Kinoshita, and F. J. Wiebel. 1972. Microsomal hydroxylases: Induction and role in polycyclic hydrocarbon carcinogenesis and toxicity. Fed. Proc. **31**: 1298–1309.

Gerber, P. 1966. Studies on the transfer of subviral infectivity from SV_{40} induced hamster tumor cells to indicator cells. Virology **28**:501–509.

German, J. 1972. Genes which increase chromosomal instability in somatic cells and predispose to cancer. Progr. Med. Genet. **8**:61–101.

German, J. (ed.). 1974. Chromosomes and Cancer. New York: John Wiley & Sons.

Gershon, D., and L. Sachs. 1963. Properties of a somatic hybrid between mouse cells with different genotypes. Nature **198**:912–913.

Gey, G. O. 1941. Cytological and cultural observations on transplantable rat sarcomata produced by the inoculation of altered normal cells maintained in continuous culture. Cancer Res. **1**:737.

Gey, G. O., M. K. Gey, W. M. Frior, and W. O. Self. 1949. Cultural and cytologic studies on autologous normal and malignant cells of specific *in vitro* origin. Conversion of normal into malignant cells. Acta Univ. Intern. Contra Cancrum **6**:706–712.

Ghose, T., S. T. Norvell, A. Guclu, and A. S. MacDonald. 1975. Immunochemotherapy of human malignant melanoma with Chlorambucil-carrying antibody. Eur. J. Cancer **11**:321–326.

Ginsberg, H. S., H. G. Pereira, R. C. Valentine, and W. C. Wilcox. 1966. A proposed terminology for the adenovirus antigens and virion morphological subunits. Virology **28**:782–783.

Gold, P., and S. O. Freedman. 1965. Demonstration of tumor-specific antigens in human colonic carcinomata by immunological tolerance and absorption techniques. J. Exp. Med. **121**:439–462.

Gold, P., and S. O. Freedman. 1975. Tests for carcinoembryonic antigen. Role in diagnosis and management of cancer. JAMA **234**:190–192.

Goldberg, N. D., M. K. Haddox, E. Dunham, C. Lopez, and J. W. Hadden. 1974. The Ying Yang hypothesis of biological control: Opposing influences of cyclic GMP and cyclic AMP in the regulation of cell proliferation and other biological processes. *In* Control of Proliferation of Animal Cells (B. Clarkson and R. Baserga, eds.), Cold Spring

Harbor Conference on Cell Proliferation, pp. 609–625. New York: Cold Spring Harbor Laboratory.

Goldie, H. 1956. Growth characteristics of free tumor cells in various body fluids and tissues of the mouse. Ann. N.Y. Acad. Sci. **63**:711–719.

Goldstein, M. N. 1972. Growth and differentiation of normal and malignant sympathetic neurons *in vitro*. *In* Cell Differentiation (R. Harris, P. Allin, and D. Viza, eds.), pp. 131–137. Copenhagen: Munksgaard.

Goldstein, M. N., J. A. Burdman, and L. J. Journey. 1974. Long-term tissue cultures of neuroblastomas. II. Morphologic evidence of differentiation and maturation. J. Nat. Cancer Inst. **32**:165–174.

Golub, S. H., *et al.* 1972. Cellular reactions against Burkitt lymphoma cells. III. Effector cell activity of leukocytes stimulated *in vitro* with autochthonous cultured lymphoma cells. Int. J. Cancer **10**:157–164.

Goss, R. J. (ed.). 1972. Regulation of Organ and Tissue growth. New York: Academic Press.

Graham, C. F. 1966. The regulation of DNA synthesis and mitosis in multinucleate frog eggs. J. Cell Sci. **1**:363–374.

Gray, J. M., and G. B. Pierce. 1964. Relationship between growth rate and differentiation of melanoma *in vivo*. J. Nat. Cancer Inst. **32**:1201–1210.

Green, M. 1966. Biosynthetic modifications induced by DNA animal viruses. Ann. Rev. Microbiol. **20**:189–222.

Green, M., J. T. Parsins, M. Pina, K. Fujinaga, H. Caffier, and I. Landgraf-Leurs. 1970. Transcription of adenovirus genes in productively infected and in transformed cells. Cold Spring Harbor Symp. Quant. Biol. **35**:803–818.

Greenberg, P. L., W. C. Nichols, and S. L. Schrier. 1971. Granulopoiesis in acute myeloid leukemia and preleukemia. N. Eng. J. Med. **284**:1225–1232.

Greenblatt, M., and P. Shubik. 1968. Tumor angiogenesis: Transfilter diffusion studies in the hamster by the transparent chamber technique. J. Nat. Cancer Inst. **41**:111–124.

Greene, H. S. N. 1951. A conception of tumor autonomy based on transplantation studies: A review. Cancer Res. **11**:899–903.

Greene, H. S. N. 1957. Heterotransplantation of tumors. Ann. N.Y. Acad. of Sci. **69**:818–829.

Greene, H. S. N., and E. K. Harvey. 1964. The relationship between the dissemination of tumor cells and the distribution of metastases. Cancer Res. **29**:799–811.

Grobstein, C. 1953. Morphogenetic interaction between embryonic mouse tissues separated by a membrane filter. Nature **172**:869–871.

Grobstein, C. 1959. Differentiation of vertebrate cells. *In* The Cell (J. Brachet and A. E. Mirsky, eds.), pp. 437–496. New York: Academic Press.

Grobstein, C., and E. Zwilling. 1953. Modification of growth and differentiation of chorioallantoic grafts of chick blastoderm pieces after cultivation at a glass-clot interface. J. Exp. Zool. **122**:259–284.

Gross, L. 1951. "Spontaneous" leukemia developing in C3H mice following inoculation, in infancy, with AK-leukemic extracts, or AK-embryos. Proc. Soc. Exp. Biol. Med. 76:27–32.

Gross, L. 1953. A filterable agent, recovered from AK leukemic extracts, causing salivary gland carcinomas in C3H line. Proc. Soc. Exp. Biol. Med. 83:414–421.

Gurdon, J. B. 1963. Nuclear transplantation in amphibians and the importance of stable nuclear changes in promoting cellular differentiation. Quart. Rev. Biol. 38:54–78.

Gurdon, J. B. 1974. The Control of Gene Expression in Animal Development, pp. 1–115. Cambridge, Mass.: Harvard University Press.

Gyorkey, F., K. W. Min, I. Krisko, and P. Gyorkey. 1975. The usefulness of electron microscopy in the diagnosis of human tumors. Hum. Path. 6:421–441.

Haddow, A. 1944. Transformation of cells and viruses. Nature 154:194–199.

Hakomori, S. 1973. Glycolipids of tumor cell membranes. Adv. Cancer Res. 18:265–315.

Hall, T. 1974. A paraneoplastic syndrome. N.Y. Acad. Sci. 230:565–570.

Hall, T. C. 1974. A word of cautious dissent. Semin. Oncol. 1:429–431.

Hall, T. C. 1976. Late effects of cancer therapy. Cancer 37:1218–1225.

Hammond, E. C. 1975a. Tobacco. *In* Persons at High Risk of Cancer (J. F. Fraumeni, Jr., ed.), pp. 131–138. New York: Academic Press.

Hammond, E. C. 1975b. The epidemiological approach to the etiology of cancer. Cancer 35:652–654.

Hanafusa, H., and T. Hanafusa. 1966. Analysis of defectiveness of Rous sarcoma virus. IV. Kinetics of RSV production. Virology 28:369–378.

Harnden, D. G. 1970. The role of genetic change in neoplasia—constitutional chromosome abnormalities. *In* Genetic Concepts and Neoplasia, 23rd M.D. Anderson Symposium, pp. 31–35. Baltimore: Williams & Wilkins Co.

Harris, H. 1967. The reactivation of the red cell nucleus. J. Cell Sci. 2:23–32.

Harris, H. 1970. Nucleus and Cytoplasm. Oxford: Clarendon Press.

Harris, H. 1971. Cell fusion and the analysis of malignancy. Proc. Roy. Acad. Sci., Series B 179:1–20.

Harris, H., and J. F. Watkins. 1965. Hybrid cells derived from mouse and man: Artificial heterokaryons of mammalian cells from different species. Nature 205:640–646.

Hayflick, L., and P. S. Moorhead. 1961. The serial cultivation of human diploid cell strains. Exp. Cell Res. 25: 585–621.

Hecker, E. 1971. Isolation and characterization of the cocarcinogenic principles from croton oil. *In* Methods in Cancer Research (H. Busch, ed.), Vol. VI, pp. 439–484. New York: Academic Press.

Heidelberger, C. 1973. Chemical oncogenesis in culture. Adv. Cancer Res. 18:317–366.

Heidelberger, C. 1970. Chemical carcinogenesis, chemotherapy: Cancer's continuing core challenges. Cancer Res. 30:1549–1569.

Hellstrom, K. E., and I. Hellstrom. 1970. Immunological enhancement as studied by cell culture techniques. Ann. Rev. Microbiol. 24:373–398.

Henle, W., and G. Henle. 1974. Epstein–Barr virus and human malignancies. Cancer 34:1368–1374.

Herberman, R. B., E. B. Rosenberg, R. H. Halterman, J. L. McCoy, and B. G. Leventhal. 1972. Cellular immune reactions to human leukemia. Nat. Cancer Inst. Monogr. 35:259.

Heston, W. E. 1963. Genetics of neoplasia. *In* Methodology in Mammalian Genetics (W. J. Burdette, ed.), p. 247. San Francisco: Holden-Day.

Heston, W. E. 1965. Genetic factors in the etiology of cancer. Cancer Res. 25:1320–1326.

Heston, W. E., and G. Vlahakis. 1967. Genetic factors in mammary tumorigenesis. *In* Carcinogenesis: A Broad Critique, Twentieth Annual Symposium on Fundamental Cancer Research, pp. 347–363. Baltimore: Williams & Wilkins.

Hewetson, J. F., S. H. Golub, G. Klein, S. Singer, and E. A. Svedmyr. 1972a. Cellular reactions against Burkitt lymphoma cell. I. Colony inhibition with effector cells from patients with Burkitt's lymphoma. Int. J. Cancer 10:142–149.

Hewetson, J. F., S. H. Golub, G. Klein, S. Singer, and E. A. Svedmyr. 1972b. Cellular reactions against Burkitt lymphoma cells. II. Effector cells obtained by allogenic stimulation in mixed leukocyte culture. Int. J. Cancer 10:150–156.

Hewetson, J. F., S. H. Golub, G. Klein, S. Singer, and E. A. Svedmyr. 1972c. Cellular reactions against Burkitt lymphoma cells. III. Effector cell activity of leukocytes stimulated *in vitro* with autochthonous cultured lymphoma cells. Int. J. Cancer 10:157–164.

Higginson, J., B. Terracini, and C. Agthe. 1975. Nutrition and cancer: Ingestion of foodborne carcinogens. *In* Cancer Epidemiology and Prevention (G. T. Stewart, ed.), pp. 177–206. Springfield, Ill.: Charles C. Thomas.

Hirschhorn, K. 1976. Chromosomes and cancer. *In* Cancer and Genetics. Birth Defects: Original Article Series (D. Bergsma, ed.), Vol. XII, No. 1, pp. 113–121.

Hitotsumachi, S., Z. Rabinowitz, and L. Sachs. 1971. Chromosomal control of reversion in transformed cells. Nature 231:511–514.

Holley, R. W. 1974. Serum factors and growth control. *In* Control of Proliferation in Animal Cells (B. Clarkson and R. Baserga, eds.), pp. 13–19. Cold Spring Harbor Conference on Cell Proliferation. Cold Spring Harbor, New York: Cold Spring Harbor Laboratory.

Holtfreter, J. 1951. Some aspects of embryonic induction. Growth Supplement 10:117–152.

Holtzer, H. 1970. Proliferation and quantal cell cycles in the differentiation of muscle, cartilage, and red blood cells. *In* Control Mechanism in the Expression of Cellular Phenotypes (H. A. Padykula, ed.). London: Academic Press.

Hoover, R., and P. Cole. 1973. Temporal aspects of occupational bladder carcinogenesis. N. Eng. J. Med. **288**:1040–1043.

Houck, J. C., and A. M. Attallah. 1975. Chalones (specific and endogenous mitotic inhibitors) and cancer. *In* Cancer (I. Becker, ed.). New York: Plenum Press.

Hruban, Z., and M. Rechcigl, Jr. 1965. Fine structure of transplantable hepatomas of the rat. J. Nat. Cancer Inst. **35**:459–473.

Hsie, A. W., C. Jones, and T. T. Puck. 1971. Further changes in differentiation state accompanying the conversion of Chinese hamster cells to fibroblastic form by dibutyryl adenosine cyclic 3′:5′-monophosphate and hormones. Proc. Nat. Acad. Sci., U.S.A. **68**:1648–1652.

Huebner, R. J., D. Armstrong, M. Okuyan, P. S. Suma, and H. C. Turner. 1964. Specific complement-fixing viral antigens in hamster and guinea pig tumors induced by the Schmidt–Ruppin strain of avian sarcoma. Proc. Nat. Acad. Sci., U.S.A. **51**:742–750.

Huebner, R. J., G. J. Kelloff, P. S. Sarma, W. T. Lane, H. C. Turner, R. V. Gilden, S. Oroszlan, H. Meier, D. D. Myers, and R. L. Peters. 1970. Group-specific antigen expression during embryogenesis of the genome of the C-type RNA tumor virus: Implications for ontogenesis on oncogenesis. Proc. Nat. Acad. Sci., U.S.A., **67**:366–376.

Hueper, W. C., F. H. Wiley, and H. D. Wolfe. 1938. Experimental production of bladder tumors in dogs by administration of beta-naphthylamine. J. Industr. Hyg. **20**:46–84.

Huggins, C. B., and C. V. Hodges. 1941. Studies on prostate cancer. 1. The effect of castration of estrogen and of androgen injection on serum phosphatases in metastatic carcinoma of the prostate. Cancer Res. **1**:293–297.

Hynes, R. O. 1976. Cell surface proteins and malignant transformation. Biochem. Biophys. Acta **458**:73–107.

Ichikawa, Y. 1970. Further studies on the differentiation of a cell line of myeloid leukemia. J. Cell Physiol. **76**:175–184.

Ichikawa, Y., D. H. Pluznik, and L. Sachs. 1966. *In vitro* control of the development of macrophage and granulocyte colonies. Proc. Nat. Acad. Sci., U.S.A. **56**:488–495.

Itaya, K., S. Hakamori, and G. Klein. 1976. Long-chain neutral glycolipid and gangliosides of murine fibroblast lines and their low- and high-tumorigenic hybrids. Proc. Nat. Acad. Sci., U.S.A. **73**:1568–1571.

Jablon, S. 1975. Radiation. *In* Persons at High Risk of Cancer—An Approach to Cancer Etiology and Control (J. F. Fraumeni, Jr., ed.), pp. 151–165. New York: Academic Press.

Jackson, R., and S. Gardere. 1971. Nevoid basal cell carcinoma syndrome. Canad. Med. Assoc. J. **105**:850–859.

Jacob, F., and J. Monod. 1970. Introduction. *In* The Lactose Operon (J. F. Beckwith and D. Zipser, eds.), pp. 1–4. Cold Spring Harbor, New York: Cold Spring Harbor Laboratory.

Jamarosmanovic, A., and W. R. Loewenstein. 1968. Intercellular communication and tissue growth. III. Thyroid Cancer. J. Cell Biol. **38**:556-561.

Jami, J., C. Failly, and E. Ritz. 1973. Lack of expression of differentiation in mouse teratoma-fibroblast somatic cell hybrids. Exp. Cell Res. 76:191–199.

Jamieson, J. D., and G. E. Palade. 1968. Intracellular transport of secretory proteins in the pancreatic exocrine cell. III. Dissociation of intracellular transport from protein synthesis. J. Cell. Biol. 39:580–588.

Jensen, E. V., G. E. Block, S. Smith, K. Kyser, and E. R. DeSombre. 1972. Estrogen receptors and hormone dependency. *In* Estrogen Target Tissues and Neoplasia (T. L. Das, ed.). Chicago: University of Chicago Press.

Jensen, E. V., G. E. Block, S. Smith, and E. R. DeSombre. 1973. Hormonal dependency of breast cancer. Recent Results in Cancer Therapy 42:55–62.

Jensen, E. V., and E. R. DeSombre. 1973. Estrogen-receptor interaction. Science 182:126–184.

Jones, P. A., W. E. Laug, and W. F. Benedict. 1975. Fibrinolytic activity in a human fibrosarcoma cell line and evidence for the induction of plasminogen activator secretion during tumor formation. Cell 6: 245–252.

Jordan, S. W. 1964. Electron microscopy of hepatic regeneration. Exp. Molec. Path. 3:183–200.

Kakunaga, T. 1975. The role of cell division in the malignant transformation of mouse cells treated with 3-methylcholanthrene. Cancer Res. 35: 1637–1642.

Kaliss, N. 1969. Immunological enhancement. Int. Rev. Exp. Path. 8:241–276.

Kaplan, H. S. 1972. *In* RNA Viruses and Host Genome in Oncogenesis (P. Emmelot and P. Bentuelzer, eds.), pp. 143–154. Amsterdam: North Holland Publishing Co.

Karcher, K.-H., and K. Jentzsch. 1972. Radiobiology as the basis of radiotherapy. Prog. Pathobiol. 2:305.

Kastendieck, H., E. Altenähr, and P. Burchardt. 1973. Elektroenmikroskopische untersuchungen zur zelldifferenzierung in prostatacarcinomen. Virchow's Arch. (Pathol. Anat.) 361:241–256.

Kellerman, G., M. Luyten-Kellermann, and C. R. Shaw. 1973. Genetic variation of aryl hydrocarbon hydroxylase in human lymphocytes. Am. J. Hum. Genet. 25:327–331.

Kellerman, G., C. R. Shaw, and M. Luyten-Kellermann. 1973. Aryl hydrocarbon hydroxylase inducibility and bronchogenic carcinoma. N. Eng. J. Med. 289:934–937.

Kennaway, E. L. 1955. The identification of a carcinogenic compound in coal-tar. Brit. Med. J. 2:749–752.

Kermode, G. O. 1972. Food Additives. Sci. Am. 226:15.

Kerr, S. J., and E. Borek. 1973. Regulation of tRNA methyltransferases in normal and neoplastic tissues. Adv. Enzyme. Reg. 11:63–77.

Kersey, J. H., B. D. Spector, and R. A. Good. 1973. Immunodeficiency and cancer. Adv. Cancer Res. 18:211–230.

Kessler, I. I. 1976. Human cervical cancer as a venereal disease. Cancer Res. 36:783–791.

Ketterer, B., E. Tipping, D. Beale, J. Meuwissen, and C. M. Kay. 1975. Chemical and viral oncogenesis. Proceedings of XI International Cancer Congress, Florence, 1974. Excerpta Media Internat. Congr. Series No. 350, Vol. 2.

Khoury, G., J. C. Byrne, and M. A. Martin. 1972. Patterns of SV_{40} DNA transcription after acute infection of permissive and non-permissive cells. Proc. Nat. Acad. Sci., U.S.A., 69:1925.

Kidd, J. G. 1953. Regression of transplanted lymphomas induced *in vivo* by means of normal guinea pig serum J. Exp. Med. 98:565–606.

King, T. J., and M. A. DiBerardino. 1965. Transplantation of nuclei from frog renal adenocarcinoma. I. Development of tumor nuclear-transplant embryos. Ann. N.Y. Acad. Sci. 126:115–126.

King, T. J., and R. G. McKinnell. 1960. An attempt to determine the developmental potentialities of the cancer cell nucleus by means of transplantation. *In* Cell Physiology of Neoplasia, pp. 591–617. Austin: University of Texas Press.

Kirchheim, D., and R. Bacon. 1969. Ultrastructural studies of carcinoma of the human prostate gland. Invest. Urol. 6:611–630.

Kirchheim, D., D. Brandes, and R. L. Bacon. 1974. Fine structure and cytochemistry of human prostatic carcinoma. *In* Male Accessory Organs: Structure and Function (D. Brandes, ed.), pp. 397–423. New York: Academic Press.

Klein, G. 1975. The Epstein–Barr virus and neoplasia. N. Eng. J. Med. 293:1353–1357.

Klein, G., and H. Harris. 1972. Expression of polyoma-induced transplantation antigen PTA in hybrid cell lines. Nature New Biol. 237:163–164.

Klein, G., and E. Klein. 1956. Conversion of solid neoplasms into ascites tumors. Ann. N.Y. Acad. Sci. 63:640–661.

Kleinsmith, L. J., and G. B. Pierce. 1964. Multipotentiality of single embryonal carcinoma cells. Cancer Res. 24:1544–1551.

Knudson, A. G., L. C. Strong, and D. E. Anderson. 1973. Heredity and cancer in man. Prog. Med. Genet. 9:113–158.

Knudson, A. G., Jr. 1971. Mutation and cancer: Statistical study of retinoblastoma. Proc. Nat. Acad. Sci., U.S.A. 68:820–823.

Konigsberg, I. R. 1965. Aspects of cytodifferentiation of skeletal muscle. *In* Organogenesis (R. L. DeHaan and H. Ursprung, eds.), pp. 337–358. New York: Holt, Rinehart & Winston.

Koprowski, H., F. C. Jensen, and Z. Steplewski. 1967. Activation of production of infectious tumor virus SV_{40} in heterokaryon cultures. Proc. Nat. Acad. Sci., U.S.A. 58:127–133.

Korant, B. D. 1975. Regulation of animal virus replication by protein cleavage. *In* Proteases to Biological Control (E. Reich, D. B. Rifkin, and E. Shaw, eds.). Cold Spring Harbor, New York: Cold Spring Harbor Laboratory.

Kortewig, R. 1934. The causes and control of mammary cancer in mice. Nederl. Tijdschr. V. Geneesk **78**:221–246. As quoted by J. J. Bittner, Harvey Lectures.

Kouri, R. E., H. Ratrie, and C. E. Whitmire. 1973. Evidence of a genetic relationship between susceptibility to 3-methylcholanthrene-induced subcutaneous tumors and inducibility of aryl hydrocarbon hydroxylase. J. Nat. Cancer Inst. **51**:197–200.

Kuschner, M., and S. Laskin. 1971. Experimental models in environmental carcinogenesis. Am. J. Path. **64**:183–196.

Lampert, F. 1971. Chromosome alterations in human carcinogenesis. Adv. Cell Mol. Biol. **1**:185–212.

Lasnitzki, I. 1958. The effect of carcinogens, hormones and vitamins on organ cultures. Int. Rev. Cytol. **7**:79–121.

Leblond, C. P., and C. E. Stevens, 1948. The constant renewal of intestinal epithelium in the albino rat. Anat. Rec. **100**:357–371.

Lee, L. F., E. D. Kieff, S. L. Bachenheimer, B. Roizman, P. G. Spear, B. R. Burmester, and K. Nazerian. 1971. Size and composition of Morek's disease virus deoxyribonucleic acid. J. Virol. **7**:289–294.

Lehman, J. M., and J. R. Sheppard. 1972. Agglutinability by plant lectins increases after RNA virus transformation. Virology **49**:339–341.

Leighton, J. 1957. Contributions of tissue culture studies to an understanding of the biology of cancer: A review. Cancer Res. **17**:929–941.

Leighton, J. 1967. The Spread of Cancer. New York: Academic Press.

Levi-Montalcini, R. 1975. Nerve growth factor. Letter to Editor, Science **187**:113.

Levy, J. A. 1974. Cats and cancer. J. Am. Med. Assoc. **229**:1654–1655.

Lewison, E. F. (ed). 1976. Conference on Spontaneous Regression of Cancer. Monograph 44, National Cancer Institute.

Lieberman, M. W., R. N. Baney, R. E. Lee, S. Sell, and E. Farber. 1971. Studies on DNA repair in human lymphocytes treated with proximate carcinogens and alkylating agents. Cancer Res. **31**:1297–1306.

Lilien, J. E. 1968. Specific enhancement of cell aggregation *in vitro*. Dev. Biol. **17**:657–678.

Lilien, J. E., and A. A. Moscona. 1967. Adenovirus endocarditis in mice. Science **157**:70–72.

Lin, H. S., C. S. Lin, S. Yeh, and S. M. Tu. 1969. Fine structure of nasopharyngeal carcinoma with special reference to the anaplastic type. Cancer **23**:390–504.

Lindberg, V., and J. E. Darnell. 1970. SV_{40}-specific RNA in the nucleus and polyribosomes of transformed cells. Proc. Nat. Acad. Sci., U.S.A. **65**:1089–1096.

Lipsett, M. B. 1968. Hormonal syndromes associated with neoplasia. Adv. Metab. Disord. **3**:111–152.

Littlefield, J. W. 1964. Selection of hybrids from matings of fibroblasts *in vitro* and their presumed recombinants. Science **145**:709–710.

Littlefield, J. W. 1966. The use of drug-resistant marker to study the hybridization of the mouse fibroblasts. Exp. Cell Res. **41**:190–196.

Loeb, L. A., C. F. Springgate, and N. Battula. 1974. Errors in DNA replication as a basis of malignant changes. Cancer Res. **34**:2311–2321.

Loewenstein, W. R. 1966. Permeability of membrane junctions. Ann. N.Y. Acad. Sci. **137**:441–472.

Loewenstein, W. R. 1973. Membrane junctions in growth and differentiation. Fed. Proc. **32**:60–64.

Luse, S. 1961. Ultrastructural characteristics of normal and neoplastic cells. Prog. Exp. Tumor Res. **2**:1–35.

Lynch, H. T., and A. J. Krush. 1971. Cancer family "G" revisited: 1895–1970. Cancer **27**:1505–1511.

Lynch, H. T., A. J. Krush, A. L. Larsen, and C. W. Magnuson. 1966. Endometrial carcinoma: Multiple primary malignancies, constitutional factors, and heredity. Am. J. Med. Sci. **252**:381–390.

Lynch, H. T., A. J. Krush, R. J. Thomas, and J. Lynch. 1976. Cancer family syndrome. *In* Cancer Genetics (H. T. Lynch, ed.), pp. 355–388. Springfield, Ill.: Charles C. Thomas.

Lyon, M. F. 1968. Chromosomal and subchromosomal inactivation. Ann. Rev. Genet. **2**:31–52.

MacKenzie, C. G., and J. B. MacKenzie. 1943. Effect of sulfonamides and thioureas on the thyroid gland and basal metabolism. Endocrinology **32**:185–209.

MacKenzie, C. G., J. B. MacKenzie, and P. Beck. 1961. The effect of pH on growth, protein synthesis, and lipid-rich particles of cultured mammalian cells. J. Biophys. Biochem. Cytol. **9**:141–156.

MacPherson, I., and L. Montagnier. 1964. Agar suspension culture for the selective assay of cells transformed by polyoma virus. Virology **23**: 291–294.

Makino, S. 1956. Further evidence favoring the concept of the stem cell in ascites tumors of rats. Ann. N.Y. Acad. Sci. **63**:818–830.

Makino, S. 1975. Neoplasia. *In* Human Chromosomes, pp. 429–516. Amsterdam: North Holland Publishing Co.

Mao, P., K. Nakao, and A. Angrist. 1966. Human prostatic carcinoma: An electron microscope study. Cancer Res. **26**:955–973.

Markert, C. L. 1968. Neoplasia: A disease of cell differentiation. Cancer Res. **28**:1908–1914.

Marx, J. L. 1976. Estrogen drugs: Do they increase the risk of cancer? Science **191**:838–840, 882.

Mathé, G., J. L. Amiel, L. Schwarzenberg, M. Schneider, M. Hayat, F. DeVassal, C. Jasmin, and C. Rosenfeld. 1973. Adoptive and active immunotherapy of experimental and human leukemias, pp. 89–110. *In* Immunology and Cancer. Ottawa: University of Ottawa Press.

Mathé, G., and R. Weiner (eds.). 1974. Investigation and Stimulation of Immunity in Cancer Patients: Recent Results in Cancer Research, Vol. 47. Heidelberg: Springer-Verlag.

Mayyasi, S. A., D. L. Larson, and M. Ahmed. 1976. Practical considerations in the development of human cancer vaccine. Cancer Res. **36**: 861–864.

McAllister, R. M., M. Nicolson, M. B. Gardner, R. W. Rongey, S. Rasheed, P. S. Sarma, R. J. Huebner, M. Hatanaka, S. Oroszlan, R. W. Gilden, A. Kabigting, and L. Vernon. 1972. C-type virus released from cultured human rhabdomyosarcoma cells. Nature New Biol. **235**:3–6.

McCann, J., E. Choi, E. Yamasaki, and B. N. Ames. 1975. Detection of carcinogens as mutagens in the Salmonella/microsome test: Assay of 300 chemicals. Proc. Nat. Acad. Sci., U.S.A. **72**:5135–5139.

McClary, D. R., and A. A. Moscona. 1974. Exp. Cell Res. **87**:438–443.

McKinnell, R. G. 1972. Nuclear transfer in Xenopus and Rana compared. *In* Cell Differentiation (R. Harris, P. Allin, and D. Viza, eds.), pp. 61–64. Copenhagen: Munksgaard.

McKinnell, R. G., L. M. Steven, Jr., and D. D. Labat. 1976. Frog renal tumors are composed of stroma, vascular elements, and epithelial cells: What type of nucleus programs for tadpoles with the cloning procedure? *In* Progress in Differentiation Research (N. Müller-Bérat, ed.). New York: Elsevier.

McNutt, N., L. A. Culp, and P. H. Black. 1971. Contact-inhibited revertant cell lines isolated from SV-40 transformed cells. II. Ultrastructural study. J. Cell Biol. **50**:691–708.

McNutt, N. S., and R. S. Weinstein. 1969. Carcinoma of the cervix: Deficiency of nexus intercellular junctions. Science **165**:597–598.

Meezan, E., H. C. Wau, P. H. Black, and P. W. Robbins. 1969. Comparative studies on the carbohydrate-containing membrane components of normal and virus-transformed mouse fibroblasts. Biochemistry **8**: 2518–2524.

Meier, H. (ed.). 1963. Epizootiology of cancer in animals. Ann. N.Y. Acad. Sci. **108**:617–1326.

Melnick, J. L., E. Adam, and W. E. Rawls. 1974. The causative role of herpesvirus type 2 in cervical cancer. Cancer **34** (Suppl.):1375–1385.

Mendelsohn, M. L. 1962. Autoradiographic analysis of cell proliferation in spontaneous breast cancer of C_3H mouse. III. The growth fraction. J. Nat. Cancer Inst. **28**:1015–1029.

Merwin, R. M. and G. H. Algire. 1959. Induction of plasma cell neoplasms and fibrosarcomas in BalB/c mice carrying diffusion chambers. Proc. Soc. Exp. Biol. Med. **101**:437.

Metcalf, D., T. R. Bradley, and W. Robinson. 1967. Analysis of colonies developing *in vitro* from mouse bone marrow cells stimulated by kidney feeder layers or leukemic serum. J. Cell Physiol. **69**:93–108.

Metcalf, D., and M. A. S. Moore. 1970. Factors modifying stem cell proliferation of myelomonocytic leukemic cells *in vitro* and *in vivo*. J. Nat. Cancer Inst. **44**:801–808.

Miller, C. A., and E. M. Levine. 1972. Neuroblastoma: Synchronization of neurite outgrowth in cultures grown on collagen. Science **177**: 799–801.

Miller, J. A. 1970. Carcinogenesis by chemicals: An overview. G. H. A. Clowes Memorial Lecture. Cancer Res. **30**:559–576.

Miller, J. A., and E. C. Miller. 1971. Chemical carcinogenesis: Mechanisms and approaches to its control. J. Nat. Cancer Inst. **47**:v–xiv.

Miller, R. W. 1975. Radiation. *In* Cancer Epidemiology and Prevention (D. Schottenfeld, ed.), pp. 93–101. Springfield, Ill.: C. C. Thomas.

Mintz, B., and K. Illmensee. 1975. Proc. Nat. Acad. Sci., U.S.A. **72**:3585–3589.

Mintz, B., K. Illmensee, and J. D. Gearhart. 1975. Developmental and experimental potentialities of mouse teratocarcinoma cells from embryoid body cores. *In* Teratomas and Differentiation (M. I. Sherman and D. Solter, eds.), pp. 59–82. New York: Academic Press.

Misugi, K., N. Misugi, and W. A. Newton. 1968. Fine structural study of neuroblastoma, ganglioneuroblastoma and pheochromocytoma. Arch. path. **86**:160–170.

Mitchison, J. M. 1969. Enzyme synthesis in synchronous cultures. Science **165**:657–663.

Monahan, J. J., and R. H. Hall. 1974. Chromatin and gene regulation in eukaryotic cells at the transcriptional level. CRC Crit. Rev. Biochem. **2**:67–112.

Mondal, S., and C. Heidelberger. 1970. *In vitro* malignant transformation by methylcholanthrene of the progeny of single cells derived from the C_3H mouse prostate. Proc. Nat. Acad. Sci., U.S.A. **65**:219–225.

Mondal, S., P. T. Iype, L. M. Griesbach, and C. Heidelberger. 1970. Antigenicity of cells derived from mouse prostate cells after malignant transformation *in vitro* by carcinogenic hydrocarbons. Cancer Res. **30**:1593–1597.

Moore, D. H., J. Charney, B. Kramarsky, E. Y. Lasfargues, N. H. Sarkar, M. J. Brennan, J. H. Burrows, S. M. Sirsat, J. C. Paymaster, and A. B. Vaidya. 1971. Search for human breast cancer virus. Nature **229**:611–615.

Moore, M. A. S., N. Williams, D. Metcalf, O. M. Garson, and A. D. F. Hurdle. 1972. Control of human leukaemic cell proliferation and differentiation in agar cultures. *In* Cell Differentiation (R. Harris, P. Allin, and D. Viza, eds.), pp. 144–150. Copenhagen: Munksgaard.

Moore, R. A. 1935. The morphology of small prostatic carcinoma. J. Urol. **33**:224–234.

Morris, H. P. 1965. Studies on the development, biochemistry and biology of experimental hepatomas. Adv. Cancer Res. **9**:227–302.

Morris, H. P., and D. R. Meranze. 1974. Induction and some characteristics of "minimal deviation" and other transplantable rat hepatomas. Special Topics in Carcinogenesis. Recent Results in Cancer Research **44**:103–114.

Morton, D. L. 1972. Immunotherapy of cancer: Present status and future potential. Cancer **30**:1647–1655.

Morton, D. L., F. R. Eilber, and R. A. Malmgren. 1970. Immunologic factors in malignant melanomas, skeletal, and soft tissue sarcomas of man. Oncology **1**:242–255.

Morton, D. L., W. T. Hall, and R. A. Malmgren. 1969. Human liposarcomas: Tissue cultures containing foci of transformed cells with viral particles. Science 165:813–815.

Moscona, A. A. 1957. The development *in vitro* of chimeric aggregates of dissociated embryonic chick and mouse cells. Proc. Nat. Acad. Sci., U.S.A. 43:184–194.

Moscona, A. A. 1961. Rotation-mediated histogenetic aggregation of dissociated cells. Exp. Cell Res. 22:455–475.

Moscona, A. A. 1962. Analysis of cell recombinations in experimental synthesis of tissues *in vitro*. J. Cell. Comp. Physio. 60 (Suppl. 1):65–80.

Moscona, A. A. 1963. Studies on cell aggregation: Demonstration of materials with selective cell-binding activity. Proc. Nat. Acad. Sci., U.S.A. 49: 742–747.

Moscona, A. A. 1971. Embryonic and neoplastic cell surfaces: Availability of receptors from concanavalin A and wheat germ agglutinin. Science 171:905–907.

Moscona, M. H., and A. A. Moscona. 1963. Inhibition of adhesiveness and aggregation of dissociated cells by inhibitors of protein and RNA synthesis. Science 142:1070–1071.

Moss, F. P., and C. P. Leblond. 1970. Nature of dividing nuclei in skeletal muscle of growing rats. J. Cell Biol. 44:459–462.

Mott, D. M., P. H. Fabisch, B. P. Sani, and S. Sorof. 1974. Lack of correlation between fibrinolysis and the transformed state of cultured mammalian cells. Biochem. Biophys. Res. Comm. 61:621–627.

Mueller, G. C. 1969. Biochemical events in the animal cell cycle. Fed. Proc. 28:1780–1789.

Mulligan, R. M. 1949. Neoplasms of the Dog. Baltimore: Williams & Wilkins Co.

Murphy, D. B. and L. G. Tilney. 1974. The role of microtubules in the movement of pigment granules in Teleost Melanophores. J. Cell Biol. 61:757–779.

Nakahara, W., and F. Fukuoka. 1948. Japan Med. J. 1:271–277.

Needham, J. 1936. New advances in the chemistry and biology of organized growth. Proc. Roy. Soc. B. 29:1577–1626.

Nicolson, G. L. 1971. Difference in topology of normal and tumor cell membranes shown by different surface distributions of ferritin-conjugated concanavalin A. Nature New Biol. 233:244–246.

Nicolson, G. L., and J. L. Winkelhake. 1975. Organ specificity of blood-borne tumor metastasis determined by cell adhesion. Nature 255:230–232.

Nicolson, G. L. 1976. Surface changes associated with transformation and malignancy. Biochim. Biophys. Acta 458:1–72.

Novikoff, A. B. 1957. A transplantable rat liver tumor induced by 4-dimethylaminoazobenzene. Cancer Res. 17:1010–1027.

Novikoff, A. B. 1961. Lysosomes and related particles. *In* The Cell (J. Brachet and A. E. Mirsky, eds.), Vol. 2, pp. 423–88. New York: Academic Press.

Novikoff, A. B., E. Essner, and N. Quintana. 1964. Golgi apparatus and lysosomes. Fed. Proc. **23**:1010–1022.

Nowell, P. C. 1965. Chromosome changes in primary tumors. Prog. Exp. Tumor Res. **7**:83–103.

Nowell, P. C. 1971. Marrow chromosome studies in "preleukemia." Cancer **28**:513–518.

Nowell, P. C. 1974. Cytogenetics. *In* Cancer: A Comprehensive Treatise. Vol. 1. Etiology: Chemical and Physical Carcinogenesis (F. F. Becker, ed.), pp. 3–31. New York: Plenum Press.

Nowell, P. C., and D. A. Hungerford. 1961. Chromosome studies in human leukemia. II. Chronic granulocytic leukemia. J. Nat. Cancer Inst. **27**:1013.

Oberling, C. H., and W. Bernhard. 1961. The morphology of the cancer cells. *In* The Cell (J. Brachet and A. E. Mirsky, eds.), pp. 405–496. New York: Academic Press.

Okada, Y., and J. Tadokoro. 1962. Analysis of giant polynuclear cell formation caused by HVJ virus from Ehrlich's ascites tumor cells. Exp. Cell Res. **26**:108–118.

Olenov, J. M., and V. J. A. Fel. 1968. The antigenic structure of tumor cells as material for a study of tissue differentiation. J. Embryol. Exp. Morphol. **19**:299–309.

O'Malley, B. W., and W. T. Schrader. 1976. The receptors of steroid hormones. Sci. Am. **234**:32–43.

Orr, C. W., and S. Roseman. 1969. Intercellular adhesions. I. A quantitative assay for measuring the rate of adhesion. J. Membr. Biol. **1**:109–124.

Ossowski, L., J. P. Quigley, G. M. Kellerman, and E. Reich. 1973. Fibrinolysis associated with oncogenic transformation. J. Exp. Med. **138**:1056–1064.

Ozanne, B., and J. Sambrook. 1971. Binding of radioactively labeled concanavalin A and wheat germ agglutinin to normal and virus-transformed cells. Nature New Biol. **232**:156–160.

Padgett, B. L., D. L. Walker, G. M. ZuRhein, R. J. Eckroade, and B. H. Dessel. 1971. Cultivation of papova-like virus from human brain with progressive multifocal leucoencephalopathy. Lancet **1**:1257–1260.

Padilla, G. M., I. L. Cameron, and A. Zimmerman (eds.). 1974. Cell Cycle Controls. New York: Academic Press.

Palade, G. E., and P. Siekevitz. 1956. Pancreatic microsomes. An integrated morphological and biochemical study. J. Biophys. Biochem. Cytol. **2**:671–690.

Papaioannou, V. E., M. W. McBurney, and R. L. Gardner. 1975. Fate of teratocarcinoma cells injected into early mouse embryos. Nature **258**:70–73.

Paran, M., L. Sachs, Y. Borak, and P. Resnitzky. 1970. *In vitro* induction of granulocyte differentiation in hematopoietic cells from leukemic and nonleukemic patients. Proc. Nat. Acad. Sci., U.S.A. **67**:1542–1549.

Parsons, D. F. 1963. Mitochondrial structure: Two types of subunits on negatively stained mitochondrial membranes. Science **140**:985–987.

Paymaster, J. C., and P. Gangadharan. 1972. Epidemiology of breast cancer in India. J. Nat. Cancer Inst. **48**:1021–1024.

Payne, L. N., and R. Chubb. 1968. Studies on the nature and genetic control of an antigen in normal chick embryos which reacts in the COFAL test. J. Gen. Virol. **3**:379–391.

Penn, I., and T. E. Starzl. 1972. Malignant tumors arising *de novo* in immunosuppressed organ transplant recipients. Transplantation **14**: 407–417.

Petersen, D. F., R. A. Tobey, and E. C. Anderson. 1968. Essential biosynthetic activity in synchronized mammalian cells. *In* The Cell Cycle (G. M. Padilla, G. L. Whitson, and I. L. Cameron, eds.), pp. 341–359. New York: Academic Press.

Peyron, A. 1939. Bull. Cancer **28**:658.

Pierce, G. B. 1961. Teratocarcinoma: A Problem in Developmental Biology, Fourth Canadian Cancer Conference (R. Biggs, ed.), Vol. 4, New York: Academic Press.

Pierce, G. B. 1966. Ultrastructure of human testicular tumors. Cancer **19**: 1963–1983.

Pierce, G. B. 1967. Teratocarcinoma: Model for a developmental concept of cancer. In Current Topics in Developmental Biology, pp. 223–246. New York: Academic Press.

Pierce, G. B. 1974. Neoplasms, differentiations and mutations. Am. J. Path. **77**:103–118.

Pierce, G. B., and F. J. Dixon, Jr. 1959. Testicular teratomas. I. The demonstration of teratogenesis by metamorphosis of multipotential cells. Cancer **12**:573–583.

Pierce, G. B., F. J. Dixon, and E. L. Verney. 1960. Teratocarcinogenic and tissue-forming potentials of the cell types comprising neoplastic embryoid bodies. Lab. Invest. **9**:583–602.

Pierce, G. B., and L. D. Johnson. 1971. Differentiation and cancer. In Vitro **7**:140–145.

Pierce, G. B., P. K. Nakane, A. Martinez-Hernandez, and J. M. Ward. 1977. Ultrastructural comparison of differentiation of stem cells of adenocarcinomas of colon and breast with their normal counterparts. J. Nat. Cancer Inst. **58**:1329–1345.

Pierce, G. B., L. C. Stevens, and P. K. Nakane. 1967. Ultrastructural analysis of the early development of teratocarcinomas. J. Nat. Cancer Inst. **39**:755–773.

Pierce, G. B., and E. L. Verney. 1961. An *in vitro* and *in vivo* study of differentiation in teratocarcinomas. Cancer **14**:1017–1029.

Pierce, G. B., and C. Wallace. 1971. Differentiation of malignant to benign cells. Cancer Res. **31**:127–134.

Pike, B. L. and W. A. Robinson. 1970. Human bone marrow colony growth in agar-gel. J. Cell Physiol. **76**:77–84.

Pinkel, D. 1971. Five-year follow-up of "total therapy" of childhood lymphocytic leukemia. J. Am. Med. Assoc. **216**:648–652.

Pitot, H. C. 1960. The comparative enzymology and cell origin of rat hepatomas. II. Glutamate dehydrogenase, choline oxidase and glucose-6-phosphatase. Cancer Res. **20**:1262–1268.

Pitot, H. C., and C. Heidelberger. 1963. Metabolic regulatory circuits and carcinogenesis. Cancer Res. **23**:1694–1700.

Pitot, H. C., T. K. Shires, G. Moyer, and C. T. Garrett. 1974. Phenotypic variability as a manifestation of translational control. *In* The Molecular Biology of Cancer (H. Busch, ed.), pp. 523–534. New York: Academic Press.

Pollack, R., R. Risser, S. Conlon, and D. Rifkin. 1974. Plasminogen activator production accompanies loss of anchorage regulation in transformation of primary rat embryo cells by Simian virus SV_{40}. Proc. Nat. Acad. Sci., U.S.A. **71**:4792–4796.

Porter, K. R. 1953. Observations on a submicroscopic basophilic component of cytoplasm. J. Exp. Med. **94**:727–750.

Porter, K. R., G. J. Todaro, and V. Fonte. 1973. A scanning electron microscopic study of surface features of viral and spontaneous transformants of mouse Balb/3T3 cells. J. Cell Biol. **59**:633–642.

Potter, M. 1968. A resumé of the current status of the development of plasma-cell tumors in mice. Cancer Res. **28**:1891–1896.

Potter, M. 1973. The developmental history of the neoplastic plasma cell in mice: A brief review of recent developments. Semin. Hematol. **10**: 19–32.

Potter, V. R. 1964. Biochemical studies on minimal deviation hepatomas. *In* Cellular Control Mechanisms and Cancer (P. Emmelot and O. Mühlbock, eds.), pp. 190–210. Amsterdam: Elsevier Publ. Co.

Potter, V. R. 1964. Biochemical perspectives in cancer research. Cancer Res. **24**:1085–1098.

Prasad, K. N. 1973. Role of cyclic AMP in the differentiation of neuroblastoma cell culture. *In* The Role of Cyclic Nucleotides in Carcinogenesis (J. Schultz and H. G. Gratzner, eds.), pp. 207–237. Miami Winter Symposium, Papanicolaou Cancer Research Institute, Miami, Florida. New York: Academic Press.

Prasad, K. N., and A. W. Hsie. 1971. Morphologic differentiation of mouse neuroblastoma cells induced *in vitro* by dibutyryl adenosine 3':5'-cyclic monophosphate. Nature New Biol. **233**:141–142.

Prehn, R. T. 1971a. Neoplasia. *In* Principles of Pathobiology (M. F. LaVia, and R. B. Hill, eds.), p. 191. New York: Oxford University Press.

Prehn, R. T. 1971b. Perspectives in oncogenesis: Does immunity stimulate or inhibit neoplasia? J. Reticuloend. Soc. **10**:1–12.

Prehn, R. T. 1974. Immunomodulation of tumor growth. Am. J. Path. **77**: 119–122.

Prehn, R. T., and J. M. Main. 1957. Immunity to methylcholanthrene-induced sarcomas. J. Nat. Cancer Inst. **18**:769–778.

Prescott, D. M. 1976. The cell cycle and the control of cellular reproduction. Adv. Genet. **18**:100–177.

Purchase, H. G. 1976. Prevention of Marek's disease: A review. Cancer Res. **36**:696–700.

Raick, A. N. 1974. Cell differentiation and tumor-promoting action in skin carcinogenesis. Cancer Res. **34**:2915–2925.

Rao, P. N., and R. T. Johnson. 1970. Mammalian cell fusion. I. Studies on the regulation of DNA synthesis and mitosis. Nature **225**:159–164.

Rapin, A. M. C., and M. M. Burger. 1974. Tumor cell surfaces: General alterations detected by agglutinins. Adv. Cancer Res. **20**:1–91.

Rehn, L. 1895. Blasengeschwulste bei Fuchsin-Arbeitern. Arch. Klin. Chirurgie **50**:588–600.

Reich, E. 1974. Tumor-associated fibrinolysis. *In* Control of Proliferation in Animal Cells (B. Clarkson and R. Baserga, eds.). Cold Spring Harbor Conference on Cell Proliferation, pp. 351–355. Cold Spring Harbor, New York: Cold Spring Harbor Laboratory.

Revel, J. P., and M. J. Karnovsky. 1967. Hexagonal array of subunits in intercellular junctions of the mouse heart and liver. J. Cell Biol. **33**:67.

Revel, J. P., A. G. Yee, and A. J. Hudspeht. 1971. Gap junctions between electronically coupled cells in tissue culture and in brown fat. Proc. Nat. Acad. Sci., U.S.A. **68**:2924–2927.

Rifkin, D. B., J. N. Loeb, G. Moore, and E. Reich. 1974. Properties of plasminogen activators formed by neoplastic human cell cultures. J. Exp. Med. **139**:1317–1328.

Rigby, P. G., P. T. Pratt, R. C. Roselof, and H. M. Lemon. 1968. Genetic relationships in familial leukemia and lymphoma. Arch. Intern. Med. **121**:67–70.

Ritchie, A. C. 1970. The classification, morphology and behaviour of tumours. *In* General Pathology 4th ed. (Lord Florey, ed.), pp. 668–719. London: Lloyd-Luke Medical Books, Ltd.

Robbins, J. H., K. H. Kraemer, M. A. Lutzner, B. W. Festoff, and H. G. Coon. 1974. Xeroderma pigmentosum. Ann. Int. Med. **80**:221–248.

Roberts, S., O. Jonasson, L. Long, E. A. McGrew, R. McGrath, and W. H. Cole. 1962. Relationship of cancer cells in the circulating blood to operation. Cancer **15**:232–240.

Roblin, R., I.-N. Chou, and P. H. Black. 1975. Proteolytic enzymes, cell surface changes and viral transformation. Adv. Cancer Res. **22**:203–260.

Ronzio, R. A., and W. J. Rutter. 1973. Effects of a partially purified factor from chick embryos on macromolecular synthesis of embryonic pancreatic epithelia. Dev. Biol. **30**:307–320.

Roth, S., E. J. McGuire, and S. Roseman. 1971. Evidence for cell-surface glycosyltransferases. Their potential role in cellular recognition. J. Cell Biol. **51**:536.

Rous, P. 1910. A transmissible avian neoplasm (sarcoma of the common fowl). J. Exp. Med. **12**:696–705.

Rous, P. 1911. A sarcoma of the fowl transmissible by an agent separable from the tumor cells. J. Exp. Med. **13**:397-411.

Rous, P., and J. G. Kidd. 1941. Conditioned neoplasms and subthreshold neoplastic states. J. Exp. Med. **73**:365–390.

Rowe, W. P. 1973. Genetic factors in the natural history of murine leukemia virus infection. G. H. A. Clowes Memorial Lecture. Cancer Res. **33**: 3061–3068.

Rowley, J. D. 1973. A new consistent chromosomal abnormality in chronic myelogenous leukemia identified by quinacrine fluorescence and Giemsa staining. Nature **243**:290.

Ruben, L. N. 1956. The effects of implanting anuran cancer into regenerating adult urodele limbs. I. Simple regenerating systems. J. Morph. **98**:389–403.

Rubin, R. W., and G. D. Weiss. 1975. Direct biochemical measurements of microtubule assembly and disassembly in Chinese hamster ovary cells. The effect of intercellular contact, cold, D_2O, and N^6,O^2-dibutyryl cyclic adenosine monophosphate. J. Cell Biol. **64**:52–53.

Ruebner, M. D. 1965. Development of preneoplastic and neoplastic lesions of the liver in male rats given 0.025 percent N-2-fluorenyldiacetamide. J. Nat. Cancer Inst. **34**:697–709.

Rytömaa, T. 1973. Chalone of the granulocyte system. Nat. Cancer Inst. Monogr. **38**:143–146.

Rytömaa, T., and K. Kiviniemai. 1968. Control of DNA duplication in rat chloroleukaemia by means of the granulocytic chalone. Eur. J. Cancer **4**:595–606.

Saffiotti, U. 1971. The laboratory approach to the identification of environmental carcinogens. Proc. 9th Canad. Cancer Res. Conf. **9**:23–36.

Sambrook, J., P. A. Sharp, and W. Keller. 1972. Transcription of SV_{40}. I. Separation of the strands if SV_{40} DNA and hybridization of the separated strands to RNA extracted from lytically infected and transformed cells. J. Mol. Biol. **70**:57–71.

Sambrook, J., H. Westphal, P. R. Srinivason, and R. Dulbecco. 1968. The integrated state of viral DNA in SV_{40}-transformed cells. Proc. Nat. Acad. Sci., U.S.A. **60**:1288–1295.

Sandberg, A. A., and M. Sakurai. 1974. Chromosomes in the causation and progression of cancer and leukemia. *In* The Molecular Biology of Cancer (H. Busch, ed.), pp. 81–106. New York: Academic Press.

Sanford, K. K., W. R. Earle, E. Shelton, E. L. Schilling, E. M. Duchesne, G. D. Likely, and M. M. Becker. 1950. Production of malignancy *in vitro*. XII. Further transformations of mouse fibroblasts to sarcomatous cells. J. Nat. Cancer Inst. **11**:351–375.

Sani, B. P., D. M. Mott, V. Jasty, and S. Sorof. 1974. Properties of the principal liver target protein of a hepatic carcinogen. Cancer Res. **34**: 2476–2481.

Sariff, A. M., P. V. Danenberg, C. Heidelberger, and B. Ketterer. 1976. Separate identities of ligandin and the H-protein, a major protein to which carcinogenic hydrocarbons are covalently bound. Biochem. Biophys. Res. Comm. **70**:869–877.

Sarma, D. S. R., S. Rajalakshmi, and E. Farber. 1975. Chemical carcinogenesis: Interactions of carcinogens with nucleic acids. *In* Cancer I

(F. F. Becker, ed.), pp. 235–288. New York: Plenum Press.

Sarma, P. S., H. C. Turner, and R. J. Huebner. 1964. An avian leucosis group-specific complement fixation reaction. Application for the detection and assay of noncytopathogenic leucosis viruses. Virology 23:313–321.

Saxén, L. 1975. Transmission and spread of kidney tubule induction. *In* Extracellular Matrix Influences on Gene Expression (H. C. Slalkin and R. C. Greulich, eds.), pp. 523–529. New York: Academic Press.

Scaletta, L. J., and B. Ephrussi. 1965. Hybridization of normal and neoplastic cells *in vitro*. Nature 205:1169.

Schabel, F. M., Jr. 1975. Concepts for systemic treatment of micrometastases. Cancer 35:15–24.

Schapira, F. 1973. Isozymes and cancer. Adv. Cancer Res. 18:77–153.

Scher, W., H. D. Preisler, and C. Friend. 1973. Hemoglobin synthesis in murine virus-induced leukemic cells *in vitro*. Effects of 5-bromo-2′-deoxyuridine, dimethylformamide and dimethylsulfoxide. J. Cell Physiol. 81:63–70.

Schnebli, H. P., and M. M. Burger. 1972. Selective inhibition of growth of transformed cells by protease inhibitors. Proc. Nat. Acad. Sci., U.S.A. 69:3825–3827.

Schottenfeld, D. 1975. Cancer Epidemiology and Prevention (G. T. Stewart, ed.). Springfield, Ill.: C. C. Thomas.

Schwartz, R. S. 1975. Another look at immunologic surveillance. N. Eng. J. Med. 293:181–184.

Seeds, N. W., A. G. Gilman, T. Amano, and M. W. Nirenberg. 1970. Regulation of axon formation by clonal lines of a neural tumor. Proc. Nat. Acad. Sci., U.S.A. 66:160–167.

Seidman, H., E. Silverberg, and A. I. Holleb. 1976. Cancer statistics, 1976. CA–A Cancer Journal for Clinicians 26:2.

Seilern-Aspang, F., and K. Kratochwil. 1962. Induction and differentiation of an epithelial tumor in the newt (*Triturus cristatus*). J. Embryol. Exp. Morphol. 10:337–356.

Selikoff, I. H., E. C. Hammond, and J. Churg. 1971. Neoplasia risk associated with occupational exposure to airborne inorganic fibers. *In* Oncology 1970, Vol. V, pp. 55–62. Chicago: Yearbook.

Sell, S. 1975. Immunology, Immunopathology and Immunity, 2nd ed. New York: Harper & Row.

Seydel, H. G. 1975. The risk of tumor induction in man following medical irradiation for malignant neoplasm. Cancer 35:1641–1645.

Shank, R. C., and P. N. Magee. 1967. Similarities between the biochemical actions of cycasin and dimethylnitrosamine. Biochem. J. 105:521–527.

Shapley, D. 1976. Nitrosamines: Scientists on the trail of prime suspect in urban cancer. Science 191:268–270.

Shapot, V. S. 1972. Some biochemical aspects of the relationship between the tumor and the host. Adv. Cancer Res. 15:253–286.

Sharon, N., and H. Lis. 1972. Lectins: Cell-agglutinating and sugar-specific proteins. Science 177:949–959.

Shearer, R. W. 1971. DNA of rat hepatomas: Search for gene amplification. Biochem. Biophys. Res. Commun. 43:1324–1328.

Sheppard, J. R. and J. M. Lehman. 1933. Cellular cyclic AMP levels and transformation by oncogenic viruses. *In* Prostaglandins and Cyclic AMP: Biological Actions and Chemical Applications (R. Kahn, ed.), pp. 261–273. New York: Academic Press.

Sheridan, J, D. 1970. Low resistance junctions between cancer cells in various solid tumors. J. Cell Biol. 45:91–99.

Shimkin, M. B. 1975. Some historical landmarks in cancer epidemiology. *In* Cancer Epidemiology and Prevention (D. Schottenfeld, ed.), pp. 60–75. Springfield, Ill.: C. C. Thomas.

Shope, R. E. 1932. A filtrable virus causing a tumor-like condition in rabbits and its relationship to virus myxomatosum. J. Exp. Med. 56:803–822.

Shope, R. E. 1933. Infectious papillomatosis of rabbits. J. Exp. Med. 58:607–624.

Shope, R. E. 1937. Immunization of rabbits to infectious papillomatosis. J. Exp. Med. 65:219–231.

Sibley, C., U. Gehring, H. Bourne, and G. M. Tomkins. 1974. Hormonal control of cellular growth. *In* Control of Proliferation in Animal Cells (B. Clarkson and R. Baserga, eds.). Cold Spring Harbor Conference on Cell Proliferation, pp. 115–124. Cold Spring Harbor, New York: Cold Spring Harbor Laboratory.

Silagi, S. 1967. Hybridization of a malignant melanoma cell line with L-cells *in vitro*. Cancer Res. 27:1953–1960.

Siminovitch, L. and A. A. Axelrad. 1963. Cell-cell interaction *in vitro*: Their relation to differentiation and carcinogenesis. *In*: Proceedings of the 5th Canadian Cancer Research Conference, Vol. 5, pp. 149–165. New York: Academic Press.

Simnett, J. D. 1974. Nuclear differentiation in the development of normal and neoplastic tissues. *In* Neoplasia and Cell Differentiation (G. V. Sherbet, ed.), pp. 1–26. London: S. Karger.

Singer, S. J., and G. L. Nicolson. 1972. The fluid mosaic model of the structure of cell membranes. Science 175:720–731.

Sivak, A., B. T. Mossman, and B. L. Van Durren. 1972. Activation of cell membrane enzymes in the stimulation of cell division. Biochem. Biophys. Res. Commun. 46:605–609.

Skipper, H. E. 1973. Successes and failures at the preclinical level. Where now? Proceedings of the 7th National Cancer Conference, pp. 109–121.

Smith, P. G., and M. C. Pike. 1976. Current epidemiological evidence for transmission of Hodgkin's disease. Cancer Res. 36:660–663.

Snell, G. E. (ed.) 1941. Biology of Laboratory Mouse. Roscoe B. Jackson Memorial Laboratory, by the Staff, Bar Harbor, Maine. New York: Dover Publications.

Solomon, A. 1976. Bence-Jones proteins and light chains of immunoglobulins. N. Eng. J. Med. 294:17–23, 91–98.

Sonneborn, T. M. 1959. Kappa and related particles in paramecium. Adv. Viruses Res. **6**:229–356.

Spemann, H. 1962. Embryonic Development and Induction. New York: Hafner.

Spiegelman, S., R. Axel, and J. Schlom. 1972. Virus related RNA in human and mouse mammary tumors. J. Nat. Cancer Inst. **48**:1205–1211.

Sporn, M. B., and C. W. Dingman. 1966. Studies on chromatin. II. Effects of carcinogens and hormones on rat liver chromatin. Cancer Res. **26**:1488.

Stein, G. S., T. C. Spelsberg, and L. J. Kleinsmith. 1974. Nonhistone chromosomal proteins and gene regulation. Science **183**:817–824.

Stevens, L. C. 1959. Embryology of testicular teratomas in strain 129 mice. J. Nat. Cancer Inst. **23**:1249–1295.

Stevens, L. C. 1962. Testicular teratomas in fetal mice. J. Nat. Cancer Inst. **28**:247–256.

Stevens, L. C. 1967. Origin of testicular teratomas from primordial germ cells in mice. J. Nat. Cancer Inst. **38**:549-552.

Stevens, L. C. 1968. The development of teratomas from intratesticular grafts of tubal mouse eggs. J. Embryol. Exp. Morph. **20**:329–341.

Stevens, L. C., Jr., and C. C. Little. 1954. Spontaneous testicular teratomas in an inbred strain of mice. Proc. Nat. Acad. Sci., U.S.A. **40**:1080–1087.

Stewart, S. E. 1953. Leukemia in mice produced by a filterable agent present in AKR leukemic tissues with notes on a sarcoma produced by the same agent. Anat. Rec. **117**:532.

Stewart, S. E., B. E. Eddy, and N. Borgese. Neoplasms in mice inoculated with a tumor agent carried in tissue culture. J. Nat. Cancer Inst. **20**:1223–1243.

Stewart, S. E., B. E. Eddy, A. M. Gochenour, N. G. Borgese, and G. E. Grubbs. 1957. The induction of neoplasms with a substance released from mouse tumors by tissue culture. Virology **3**:380–400.

Stewart, S. E., G. Kasnic, C. Draycott, W. Feller, A. Golden, E. Mitchell, and T. Ben. 1972. Activation *in vitro* by 5-iododeoxyuridine of a latent virus resembling C-type virus in a human sarcoma cell line. J. Nat. Cancer Inst. **48**:273–277.

Stewart, S. E., J. Landon, and E. Lovelace. 1964. Viruses in cultures of human leukemic cells. *In* Att. Del Simposio Sul. Tena. Virus nelle leucemie dei mammiferi. Acad. Nat. dei Lincei, Rome. **364**:271.

Stoker, M., and I. MacPherson. 1961. Studies on transformation of hamster cells by polyoma virus *in vitro*. Virology **14**:359–370.

Stone, W. H., J. Friedman, and A. Fregin. 1964. Possible somatic cell mating in twin cattle with erythrocyte mosaicism. Proc. Nat. Acad. Sci., U.S.A. **51**:1036–1044.

Sugarbaker, E. V., A. M. Cohen, and A. S. Ketcham. 1971. Do metastases metastasize? Ann. Surg. **174**:161–166.

Sutherland, E. W. 1972. Studies on the mechanism of hormone action. Science **177**:401–408.

Svoboda, D., and J. Higginson. 1968. A comparison of ultrastructural changes in rat liver due to chemical carcinogens. Cancer Res. 28:1703–1733.

Sweet, B. H., and M. R. Hilleman. 1960. The vacuolating virus, SV_{40}, in liver polio vaccine. Proc. Soc. Exp. Biol. Med. 105:420–427.

Swenson, D. H., J. A. Miller, and E. C. Miller. 1973. 2,3-Dihydro-2,3-dihydroxyaflatoxin B_1; An acid hydrolysis product of an RNA aflatoxin B_1 adduct formed by hamster and rat liver microsomes *in vitro*. Biochem. Biophys. Res. Commun. 53:1260–1267.

Taylor, A. I. 1970. Dq-, Dr and retinoblastoma. Humangenetik 10:209–217.

Taylor, J. M., S. Cohen, and W. M. Mitchell. 1970. Epidermal growth factor: High and low molecular weight forms. Proc. Nat. Acad. Sci., U.S.A. 67:164–171.

Tegtmeyer, P. 1972. Simian virus 40 deoxyribonucleic acid synthesis: The viral replicon. J. Virol. 10:591–598.

Temin, H. M., and S. Mizutani. 1970. RNA-dependent DNA polymerase in virions of Rous sarcoma virus. Nature 226:1211–1213.

Temin, H. M., G. L. Smith, and N. C. Dular. 1974. Control of multiplication of normal and Rous sarcoma virus-transformed chicken embryo fibroblasts by purified multiplication stimulating activity with nonsuppressible insulin-like and sulfaction factor activities. *In* Control of Proliferation in Animal Cells (B. Clarkson and R. Baserga, eds.), pp. 19–26. Cold Spring Harbor, New York: Cold Spring Harbor Laboratory.

Thomas, L. 1959. Cellular and Humoral Aspects of the Hypersensitive States (H. S. Lawrence, ed.), pp. 529–532. New York: Hoeber-Harper.

Till, J. E., and E. A. McCulloch. 1961. A direct measurement of the radiation sensitivity of normal mouse bone marrow cells. Radiat. Res. 14:213–222.

Tobey, R. A., L. R. Gurley, C. E. Hildebrand, R. L. Ratliff, and R. A. Walters. 1974. Sequential biochemical events in preparation for DNA replication and mitosis. *In* Control of Proliferation in Animal Cells (B. Clarkson and R. Baserga, eds.), pp. 665–679. Cold Spring Harbor, New York: Cold Spring Harbor Laboratory.

Todaro, G. J., and R. J. Huebner. 1972. The viral oncogene hypothesis: New evidence. Proc. Nat. Acad. Sci., U.S.A. 69:1009–1015.

Toker, C., and N. Trevino. 1966. Ultrastructure of human primary hepatic carcinoma. Cancer 19:1594–1606.

Tokuhata, G. K. 1964. Familial factors in human lung cancer and smoking. Am. J. Pub. Health 54:24–32.

Tolmach, L. J., B. G. Weiss, and L. E. Hopwood. 1971. Ionizing radiations and the cell cycle. Fed. Proc. 30:1742–1751.

Tomashefsky, P., J. Furth, J. K. Lattimer, J. Tannenbaum, and J. Priestley. 1972. The Furth-Columbia rat Wilms' tumor. J. Urol. 107:348–354.

Tooze, J. (ed.). 1973. The Molecular Biology of Tumor Viruses. Cold Spring Harbor, New York: Cold Spring Harbor Laboratory.

Townes, P. L., and J. Holtfreter. 1955. Directed movements and selective adhesion of embryonic amphibian cells. J. Exp. Zool. 128:53–120.

Trentin, J. J., Y. Yabe, and G. Taylor. 1962. The quest for human cancer viruses. Science 137:835–841.

Trinkaus, J. P. (In preparation). Cells into Organs, 2nd ed. Prentice-Hall Inc. Englewood Cliffs, New Jersey.

Troll, W., A. Klassen, and A. Janoff. 1970. Tumorigenesis in mouse skin: Inhibition by synthetic inhibitors of proteases. Science 169:1211–1213.

Trouet, A., D. Campeneere, and C. Seduve. 1972. Chemotherapy through lysosomes with a DNA-daunorubicin complex. Nature New Biol. 239:110–112.

Umbreit, J., and S. Roseman. 1975. A requirement for reversible binding between aggregating embryonic cells before stable adhesions. J. Biol. Chem. 250:9360–9368.

Ursprung, H., and C. L. Markert. 1963. Chromosome complements of *Rana pipiens* embryos developing from eggs injected with protein from adult liver cells. Dev. Biol. 8:309–321.

Valeriote, F., and L. Van Putten. 1975. Proliferation-dependent cytotoxicity of anticancer agents: A review. Cancer Res. 35:2619–2630.

Vanky, F., J. Stjernsward, and U. Nilsonne. 1971. Cellular immunity to human sarcoma. J. Nat. Cancer Inst. 46:1145–1151.

Van Putten, L. M. 1974. Are cell kinetic data relevant for the design of tumour chemotherapy schedules? Cell Tiss. Kinet. 7:493–504.

Visfeldt, J. 1963. Transformation of sympathicoblastoma into ganglioneuroma with a case report. Acta Path. Microbiol. Scand. 58:414.

Voorhees, J. J., E. A. Duell, W. H. Kelsey, and E. Hayes. 1972. Effects of alpha and beta adrenergic stimulation on cyclic AMP formation and mitosis in epidermis. Cancer Res. 20:419.

Waddington, C. H. 1935. Cancer and the theory of organizers. Nature 135:606–608.

Walker, P. R., and V. R. Potter. 1972. Isozyme studies on adult, regenerating, precancerous and developing liver in relation to findings in hepatomas. Adv. Enz. Reg. 10:339–362.

Wallace, A. C. 1961. Metastasis as an aspect of cell behavior. Canad. Cancer Conf. 4:139–165.

Wang, L.-H., P. Duesberg, P. Mellon, and P. K. Vogt. 1976. Distribution of envelope-specific and sarcoma-specific nucleotide sequences from different parents in the RNAs of avian tumor virus recombinants. Proc. Nat. Acad. Sci., U.S.A. 73:1073–1077.

Warburg, O. (ed.). 1930. The Metabolism of Tumors. London: Constable Press.

Warren, L., J. P. Fuhrer, and C. A. Buck. 1972. Surface glycoproteins of normal and transformed cells. A difference determined by sialic acid and a growth-dependent sialyl transferase. Proc. Nat. Acad. Sci., U.S.A. 69:1838.

Warthin, A. S. 1913. Heredity with reference to carcinoma as shown by the study of the cases examined in the pathological laboratory of the University of Michigan, 1895–1913. Arch. Intern. Med. 12:546–555.

Warwick, G. P. 1971. Effect of the cell cycle on carcinogenesis. Fed. Proc. 30: 1760–1767.

Watkins, J. F., and R. Dulbecco. 1967. Production of SV_{40} virus in hetero-karyons of transformed and susceptible cells. Proc. Nat. Acad. Sci., U.S.A. 58:1396–1403.

Weber, G. 1973. Ordered and specific pattern of gene expression in neoplasia. Adv. Enz. Reg. 11:78–102.

Weber, K., E. Lazarides, R. D. Goldman, A. Vogel, and R. Pollack. 1974. Localization and distribution of action fibers in normal, transformed and revertant cells. Cold Spring Harbor Symp. xxxix:363–369.

Wegman, D. H., J. M. Peters, R. F. Jaeger, W. A. Burgess, and L. I. Boden. 1976. Vinyl chloride: Can the worker be protected? N. Eng. J. Med. 294:563–657.

Weiner, L. P., R. M. Herdon, O. Marayan, R. T. Johnson, K. Shaht, L. J. Rubinstein, T. J. Preziosi, and F. K. Conley. 1972. Isolation of virus related SV_{40} from patients with progressive multifocal leukocephalopathy. N. Eng. J. Med. 286:385–390.

Weinstein, I. B. 1970. Modifications in transfer RNA during chemical carcinogenesis. *In* Genetic Concepts of Neoplasia, pp. 380–408. The University of Texas M. D. Anderson Hospital and Tumor Institute, 23rd Annual Symposium on Fundamental Cancer Research, 1969. Baltimore: Williams & Wilkins Co.

Weinstein, I. B., N. Yamaguchi, R. Gebert, and M. E. Kaighn. 1975. Use of epithelial cell cultures for studies on the mechanisms of transformation by chemical carcinogens. In Vitro 11:130–141.

Weiss, L., and T. S. Hauschka. 1970. Malignancy, electrophoretic mobilities and sialic acids at the electrokinetic surface of TA 3 cells. Int. J. Cancer 6:270–274.

Weiss, P. 1950. Perspectives in the field of morphogenesis. Quant. Rev. Biol. 25:177–198.

Weiss, P., and G. Andres. 1952. Experiments on the fate of embryonic cells (chick) disseminated by the vascular route. J. Exp. Zool. 121:449–487.

Weiss, P., and J. L. Kavanau. 1957. A model of growth and growth control in mathematical terms. J. Gen. Physiol. 41:1–48.

Wenner, C. E. 1975. Regulation of energy metabolism in normal and tumor tissues. *In* Cancer (F. F. Becker, ed.), Vol. 3. New York: Plenum Press.

Wessels, N. K. 1963. Effects of extra-epithelial factors on the incorporation of thymidine by embryonic epidermis. Exp. Cell Res. 30:36–55.

Weston, J. A. 1971. Neural crest cell migration and differentiation. *In* Cellular Aspects of Growth and Differentiation in Nervous Tissue (D. Pease, ed.). UCLA Forum in Medical Sciences 14:1.

Whang, J., E. Frei, J. H. Tjio, P. P. Carbone, and G. Brecher. 1963. The distribution of the Philadelphia chromosome in patients with chronic myelogenous leukemia. Blood J. Hematol. 22:664–673.

Whisson, M. E. 1972. The plasmogene deletion theory and the appearance of "new" gene products. *In* Cell Differentiation (R. Harris, P. Allin, and D. Viza, eds.), pp. 138–143. Copenhagen: Munksgaard.

White, D. C. 1976. The histopathologic basis for functional decrements in late radiation injury in diverse organs. Cancer 37:1126–1143.

White, P. R., and A. C. Braun. 1942. A cancerous neoplasm of plants. Autonomous bacteria-free crown-gall tissue. Cancer Res. 2:597–617.

Wiener, F., G. Klein, and H. Harris. 1971. The analysis of malignancy by cell fusion. III. Hybrids between diploid fibroblasts and other tumor cells. J. Cell Sci. 8:681–692.

Wiener, F., G. Klein, and H. Harris. 1974. The analysis of malignancy by cell fusion. VI. Hybrids between different tumor cells. J. Cell Sci. 16:189–198.

Wigler, M., and I. B. Weinstein. 1976. Tumor promoter induces plasminogen activator. Nature 259:232–233.

Wile, U. J., and L. B. Kingery. 1919. The etiology of common warts. J. Am. Med. Assoc. 73:970–973.

Willis, R. A. 1967. Pathology of Tumours, 4th ed. London: Butterworth & Co.

Willis, R. A. 1973. The Spread of Tumours in the Human Body, 3rd ed. London: Butterworth & Co.

Wilson, B., M. A. Lea, G. Vidali, and V. G. Allfrey. 1975. Fractionation of nuclei and analysis of nuclear proteins of rat liver and Morris hepatoma 7777. Cancer Res. 35:2954–2958.

Wilson, H. N. 1907. On some phenomena of coalescence and regeneration in sponges. J. Exptl. Zool. 5:245–258.

Wilson, P. D., and L. M. Franks. 1972. The ultrastructure of tumours derived from spontaneously transformed tissue culture cells. Br. J. Cancer 26:380–387.

Winocour, E., A. M. Kaye, and V. Stollar. 1965. Synthesis and transmethylation of DNA in polyoma-infected cultures. Virology 27:156–169.

Wolfe, L. G., L. A. Falk, and F. Deinhardt. 1971. Oncogenicity of Herpesvirus samiri in marmoset monkeys. J. Nat. Cancer Inst. 47:1145–1162.

Wood, S., Jr. 1958. Pathogenesis of metastasis formation observed in the rabbit ear chamber. AMA Arch. Path. 66:550–568.

Wood, S., Jr., E. D. Holyoke, and J. H. Yardley. 1961. Mechanisms of metastasis production by bloodborne cancer cells. Canad. Cancer Conf. 4:167–223.

Wu, H. C., E. Meezan, P. H. Black, and P. W. Robbins. 1969. Comparative studies on the carbohydrate-containing membrane components of normal and virus transformed mouse fibroblasts. I. Glucosamine-labeling patterns in 3T3, spontaneously transformed 3T3, and SV_{40} transformed 3T3 cells. Biochemistry 8:2509–2517.

Wurster-Hill, D. H. 1975. Chromosome banding and its application to cancer research. *In* Methods in Cancer Research (H. Busch, ed.), Vol. XI, pp. 1–41. New York: Academic Press.

Wylie, C. V., P. K. Nakane, and G. B. Pierce. 1973. Degrees of differentiation in nonproliferating cells of mammary carcinoma. Differentiation 1:11–20.

Wynder, E. L., and D. Hoffman. 1976. Tobacco and tobacco smoke. Semin. Oncol. 3:5–15.

Wynder, E. L., and B. Reddy. 1973. Studies of large-bowel cancer: Human leads to experimental application. J. Nat. Cancer Inst. 50:1099–1106.

Yahara, I., and G. M. Edelman. 1975. Electron microscopic analysis of the modulation of lymphocyte receptor mobility. Exp. Cell. Res. 91:125–142.

Yamada, K., N. Takagi, and A. A. Sandberg. 1966. Chromosomes and causation of human cancer and leukemia. II. Karyotypes of human solid tumors. Cancer 19:1879–1890.

Yamagiwa, K., and K. Ichikawa. 1918. Experimental study of the pathogenesis of carcinoma. J. Cancer Res. 3:1–21.

Yoeman, L. C., C. W. Taylor, J. J. Jordan, and H. Busch. 1975. Differences in chromatin proteins of growing and nongrowing tissues. Exp. Cell Res. 91:207–215.

Yokoyama, M., K. Okada, A. Tokue, H. Takayasu, and R. Yamada. 1971. Ultrastructural and biochemical study of neuroblastoma and ganglioneuroblastoma. Invest. Biol. 9:156–164.

Zamchek, N. 1975. The present status of CEA in diagnosis, prognosis and evaluation of therapy. Cancer 36:2460–2468.

Zamcheck, N. and H. Z. Kupchik. 1974. The interdependence of early clinical investigations and methodological development in the evolution of assays for carcinoembryonic antigen. Cancer Res. 34:2131–2136.

Zeidman, I. 1957. Metastasis: A review of recent advances. Cancer Res: 17:157–162.

Zeidman, I. 1965. Fate of circulating tumor cells. III. Comparison of metastatic growth produced by tumor cell emboli in veins and lymphatics. Cancer Res. 25:324–327.

Zubrod, C. G. 1976. Present accomplishments of cancer chemotherapy. *In* Progress in Clinical Cancer (I. M. Ariel, ed.), pp. 73–77. New York: Grune & Stratton.

Zur Hausen, H., and H. Schulte-Holthausen. 1970. Presence of E. B. virus nucleic acid homology in a "virus-free" line of Burkitt tumour cells. Nature 227:245–248.

Index

Neoplasia *(cont'd)*
behavior, 3–7
benign *(see* Benign neoplasms)
biochemistry, 58–67 *(see also*
Biochemistry)
cancer, 8 *(see also* Malignant
neoplasms)
carcinoma *(see* Carcinoma)
chromosomes, 118–121 *(see also*
Chromosomes)
classification, 7–12
definition, 3
development, 14–26
differentiation *(see* Differentiation)
effects on host, 5–7
examination, 9
features, in non-neoplastic systems,
127–128
genetics, 115–119 *(see also* Inheritance,
cancer)
growth, 3, 15 *(see also* Cell cycle;
Growth; Tissue renewal)
heterogeneity in tissue, 24, 44–47, 50–
51
histology, 50–52
immunology, 152–158 *(see also*
Antigens, tumor; Immunology)
inheritance, 115–119 *(see also*
Inheritance, cancer)
malignant *(see* Malignant neoplasms)
origin, 14–26
pathology, 3–7, 12
prognosis, 12, 168
properties, 3–7
senescence, 52
spontaneous regression, 180–182
stability, 25, 123–124, 127–128, 130–137
therapy, 172–182 *(see also* Chemotherapy; Immunotherapy;
Radiation, therapy)
ultrastructure, 43–47, 52–57
Nephroblastoma, 8 *(see also* Wilms'
tumor)
Nerve Growth Factor, 144–145
Neuraminidase, 179
Neuroblastoma, 38–40, 181, 187–190
Neurofibromatosis, 118
Nevoid basal cell carcinoma syndrome,
117–118
Nevus, 198–199
Nexus, 164
Nitrogen mustard, 175
Nitrosamines, 111
Nitrosoureas, 175
Nomenclature, neoplasms, 7–8
Non-histone acidic nucleoproteins, 137–
138
Nucleic acid, hybridization, 105, 107
Nucleic acid, metabolism, 59–60, 137–140
(see also DNA; Methylases; RNA)
Nucleoproteins, 60–61, 137–138
Nucleus, transplantation, 130–133
Nude mouse, 152

Oncogene, 130 *(see also* Integration,
viral)
Oncornavirus, 97, 108–109
Ovalbumin, 138

Papanicolaou test, 195
Papovaviruses, 104 *(see also* Advenovirus;
Herpes virus; Polyoma)
Paraproteins, 197
Parsi, 119, 191
Phase-specific antigens, 158
Philadelphia chromosome, 121
Phorbol esters, 16–17
Plants, neoplasms, 34–35
Plasminogen activator, 61–62
Polyoma, 85, 95, 103–106, 150, 154
Progesterone, 138–139
Prognosis, 12, 168
Progression, 23–26, 87, 165
Prolactin, 176
Promoters, 15–18, 125
Promotion *(see* Promoters)
Prostate:
carcinoma, 19, 21, 45, 160
hyperplasia, 21, 45, 71
Protease inhibitors, 62
Protein synthesis, ectopic, 67, 128
Proteolytic enzymes, 61–62
Pseudodiploid, 121

Radiation:
carcinogenesis, 109, 112, 126
therapy, 173–174
RD-114, 97
Red dye, 110
Regeneration, 70, 148
Renewal, cell and tissue, 52, 69–74, 84–
87
Repression, of genome, 128–129, 141
(see also Controls;
Heterochromatin)
Repressors, growth, 142–144
Rescue, viral, 105
Retinoblastoma, 117, 129
Reverse transcriptase, 97, 108
Reversibility of growth, 71–72, 78
Revertants, 136
RNA:
metabolism, 60, 137–140
viruses, 57, 140–141
RNA-dependent DNA polymerase, 97,
108
Rous Sarcoma Virus, 93, 97, 140

Sarcoma, 8 *(see also* RD-114)
Saturation density, 163
Selection *(see also* Progression)
malignant cells, 136–137, 170
transformed cells, 100
Seminoma, 43–44
Senescence, neoplasms, 52
Septicemia, 7
Serine dehydratase, 139
Serum, effects on growth, 145–146, 167